ECOLOGICAL MIGRATION

MASAYOSHI NAKAWO, YUKI KONAGAYA & SHINJILT
(EDS.)

ECOLOGICAL MIGRATION

ENVIRONMENTAL POLICY IN CHINA

PETER LANG
Bern · Berlin · Bruxelles · Frankfurt am Main · New York · Oxford · Wien

Bibliographic information published by Die Deutsche Nationalbibliothek
Die Deutsche Nationalbibliothek lists this publication in the Deutsche National-
bibliografie; detailed bibliographic data is available on the Internet at
‹http://dnb.d-nb.de›.

British Library and Library of Congress Cataloguing-in-Publication Data:
A catalogue record for this book is available from *The British Library*,
Great Britain.

Library of Congress Cataloging-in-Publication Data

Ecological migration : environmental policy in China / Masayoshi Nakawo,
Yuki Konagaya, Shinjilt, [editors].
p. cm.
Includes index.
ISBN 978-3-0343-0343-9
1. Human geography–China. 2. Human ecology–China. 3. Human beings–China–
Migations. 4. Forced migration–China. 5. Environmental policy–China.
6. China–Emigration and immigration. 7. China–Politics and government.
8. China–Environmental conditions. I. Nakao, Masayoshi. II. Konagaya, Yuki, 1957-
GF656.E43 2010
333.71'40951–dc22

2010053958

Cover illustration: © by Masayoshi Nakawo
Cover design: Eva Rolli, Peter Lang AG

ISBN 978-3-0343-0343-9

© Peter Lang AG, International Academic Publishers, Bern 2010
Hochfeldstrasse 32, CH-3012 Bern, Switzerland
info@peterlang.com, www.peterlang.com, www.peterlang.net

All rights reserved.
All parts of this publication are protected by copyright.
Any utilisation outside the strict limits of the copyright law, without the permission
of the publisher, is forbidden and liable to prosecution.
This applies in particular to reproductions, translations, microfilming, and storage
and processing in electronic retrieval systems.

Printed in Switzerland

Preface to English Edition

In 2002, the Chinese government initiated the systematic implementation of an "ecological migration" policy in the Heihe River Basin in western China. At that time, we happened to be carrying out an interdisciplinary research project entitled "Historical evolution of adaptability in an oasis region to water resource changes" in the same basin. Examining changes in the water circulation system that resulted from the Chinese government's implementation of this policy, we came to realize that the water environment is being significantly affected.

There are many reasons for and examples of migrations in the world. Their root causes can be economic factors such as poverty, social factors such as education, and, in many cases, environmental threats (Renaud et al., 2007). The cause of the migration we have witnessed in the Heihe River Basin was the implementation of the abovementioned government policy which tried to maintain or restore a "good" environment, without giving thought to the migrants themselves. It seems to have been unique in this sense.

For this reason, in 2004, we held a symposium with regard to the ecological migration taking place in the basin, in an effort to assess the ecological, economical, and cultural aspects of the migration policy. Many of the papers presented at the symposium revealed the effects of the implementation of the policy in the Heihe River Basin, since we were able to examine the situation *in situ*, within the framework of our own project. A summary of the symposium was first published in Japanese, in 2005, and translated into and published in Chinese later that same year. These Japanese and Chinese summaries have also been adopted for use by courses at several universities in Japan and China. Now, it is a great pleasure to have it translated into English, and published by the Peter Lang Publishing Group.

Roughly four years have passed since the Japanese edition was published, and the migration in the basin seems to have been completed on schedule. There are, however, a variety of results which are not coherent with the aims of the policy. Water shortages have worsened, for example, particularly in the lower reaches of the basin, and shallow wells have dried up as the level of the ground water has dropped. As a result, people have tended to dig deeper wells to satisfy their water demands, and this has led to a further lowering of the ground water level. Riverside trees have deteriorated, together with the grasses nearby, although herdsmen no longer keep their animals in the vicinity of the river.

In addition to this impact on the ecology, economical and cultural aspects have been affected as well. For example, many young grazing families, who left their grazing areas and received full government compensation for non-grazing activities, have become parasites, living on governmental money and abandoning work altogether. When they were not adequately accommodated, many also migrated to urban areas for additional income, causing civic problems as well. This had been predicted in one of the chapters herein four years ago, and it actually came to pass.

There is a new movement, however, by non-governmental organizations in China that work to help the local people to maintain their culture and a stable income, while preserving the environment as well. Their efforts concentrate on extensive discussions with the people, so that they will understand the goals, measures, and their interrelations. Their trials and efforts, however, must certainly be based on a real understanding, to which, I hope, this book will contribute.

Masayoshi Nakawo
May 30, 2009

Renaud, F., J.J. Bogardi, O. Dun, and K. Warner (2007): Control, adapt or flee – how to face environmental migration. InterSecTions, No. 5, Publication Series of UNU-EHS, United Nations University, p. 44.

Foreword

There is currently a surge of interest in global environmental issues around the world. Eco-friendly products, advertised as "kind to the earth," are selling well, although they tend to be expensive. Huge numbers of visitors are attending this year's World Expo in Aichi, Japan, dubbed the "Love the Earth" Expo, which takes up the theme of "Nature's Wisdom" and reflects from a global perspective on the sustainable coexistence of all forms of life on Earth.

I feel that human beings are basically considerate. As soon as they develop an awareness of environmental problems, many people switch to cars that produce fewer exhaust emissions and, to help prevent global warming, buy refrigerators that do not use fluorocarbon gases. To prevent desertification, people participate in afforestation tours of desert areas, or, if not able to participate directly, offer money to provide seedlings for such efforts. Throughout the world, with growing momentum, people are trying to do whatever they can to solve environmental problems.

One new expression which has been heard in China with increasing frequency in recent years is "ecological migration," which is a policy measure that attempts to deal with certain environmental problems. In short, it refers to the organized migration of people engaged in occupations that cause ecological destruction, and efforts to rehabilitate and conserve the affected areas. This strategy can be justified as a way of tackling environmental problems.

In the vast arid and semi-arid regions that make up the steppes of Inner Mongolia, grassland vegetation is in imminent danger of dying out due to overgrazing. That is, herders are grazing more livestock than the grasslands can sustain. Therefore, if the herders are made to migrate to other areas, the affected grasslands can be regenerated. It's a simple matter.

So, what has been the result of the strategy of "ecological migration?" Have the grasslands been conserved and regenerated? Can the desertification of Inner Mongolia be prevented?

This book is a collection of essays based on presentations made at the China-Japan Symposium on Ecological Migration, held in Beijing in July

2004, which was jointly hosted by the Research Institute for Humanity and Nature (Japan), and the Institute of Ethnology and Anthropology, Chinese Academy of Social Sciences.

Since the term "ecological migration" is still rather unfamiliar in itself, the book explores the strategy by considering the nature of ecological migration and its method of implementation, along with the places in which it has been carried out. To describe the results of ecological migration, we discuss changes in local ecosystems and economies, as well as the transformation of regional cultures.

Because of the current focus on environmental issues, many people instinctively tend to support measures that aim to address environmental problems, whatever they may be. To determine whether such measures are really effective, however, it is essential that they are academically verified. It may be found that the results are different in different areas or depending on the method of implementation. It is also possible that in some cases the outcome will be completely different from what is intended. In the context of the current wave of global environmental concern, this book academically considers measures aimed at solving environmental problems, through the example of "ecological migration."

May 2005
Masayoshi Nakawo

Contents

Introduction: Remote regions of western China
and "ecological migration"
SHINJILT 11

I *Questioning ecological aspects? Can "ecological migration" result in environmental conservation?*

1 The beginnings of "ecological migration" in the
   Heihe River valley from case studies in Ejene banner,
   Alasha League of the Inner Mongolia Autonomous Region
   YUKI KONAGAYA 43

2 The groundwater resource crisis caused by "ecological migration"
   Case studies of Mongolian pastoralists in Ejene Banner,
   Alasha League in the Inner Mongolia Autonomous Region
   KANAKO KODAMA 61

3 New round of grassland cultivation accompanying
   "ecological migration"
   From case studies of herders in Xianghuang Banner,
   Shilingol League, Inner Mongolia Autonomous Region
   SUYE 79

4 Forest restoration without reliance upon "ecological migration":
   From a case study of NGO activities in Guizhou Province
   YOSHIKI SEKI and XIANG HU 97

II *Questioning economic aspects: Can "ecological migration" achieve a reduction in poverty?*

5 The mechanism of poverty resulting from "ecological migration"
   From case studies of herders in Minghua District,
   Sunan Yogor Autonomous County, Gansu Province
   MAILISHA 121

Contents

6   The effectiveness of "ecological migration" in reducing poverty (1)
    A case study based on the Tarim River Basin, Xingjiang
    LI JINGYI 137

7   The effectiveness of "ecological migration" in reducing poverty (2)
    Lessons from the Implementation of Ecological Migration
    in Alasha League, Inner Mongolia Autonomous Region
    SHI GUOQING 159

8   The voluntariness of migration under the "ecological migration"
    policy: from case studies of herders in Ordos City,
    Inner Mongolia Autonomous Region
    SHUNJI ONIKI and B. GENSUO 185

*III Questioning cultural aspects: What kind of transformation
    does "ecological migration" effect?*

9   Cultural acceptance of inhabitants in "ecological migration"
    from case studies in Xianghuang Banner, Shilingol League,
    Inner Mongolia Autonomous Region
    ALTA 205

10 Villagers' perception of nature in relation to "ecological migration"
    A case study of "A" Village, Sunan Yogor Autonomous County,
    Gansu Province
    SHINJILT 223

11 Differences in perception among the parties concerned
    with the "ecological migration policy": from case studies
    in "B" Township, Qifeng District, Sunan Yogor
    Autonomous County, Gansu Province
    TOMOKO NAKAMURA 241

Conclusion
Global environmental problems and ecological migration
MASAYOSHI NAKAWO 257

Appendix 273

Afterword 277

Afterword to English Edition 281

Introduction: Remote regions of western China and "ecological migration"

SHINJILT

Introduction

Within the context of the current tide of globalization (*quanqiuhua* in Chinese), China is intent on national integration through policies that emphasize political stability and economic growth. The country has achieved remarkable economic development under a system of "Chinese-style socialism" that combines the principles of political one-party rule with free market economics. At the same time, China currently faces a number of problems, specifically environmental problems, such as the degradation of the natural environment caused by chronic population growth, economic problems such as those evidenced by the widening disparity in personal incomes between the "east" and the "west" of the country, and ethnic issues or national question *(minzu wenti)* associated with differences in tradition and culture, and related to subsistence patterns and lifestyles. Sustainable development in China thus depends on whether solutions can be found to these problems.

In considering the prospects of China's future development, it is impossible to ignore the agricultural development and migration issues of the country's past. Through the practice of agriculture, the area inhabited by Han Chinese has expanded to the extent that almost all of the river valleys, coastal areas and oases of China are inhabited by Han agricultural settlements today. These areas are either referred to as being "inland" or "east" and are considered to be "developed zones." The population growth of the Han brought about by agriculture has forced them to migrate outwards towards the so-called "remote regions" located at the peripheries of the developed zones. In these remote regions, the Han migrants cleared increasingly large areas of land for cultivation, even when the natural conditions were unsuitable for agriculture. It is thought that this practice is the principal cause of the environmental problems that have arisen in these remote regions. Even today, the majority of ethnic minorities inhabit the "west" and this region is currently the last frontier of negotiations between the state and ethnic groups. In an attempt to simultaneously solve the environmental, economic and ethnic problems in this region, the West Development Project was initiated.

Through this project, the Chinese government began to actively promote the development of transport and communications infrastructure and the cultivation of human resources in the west. The policy of "ecological migration" has been the focus of considerable attention, and it has become manifested as one of the most fundamental foundations of recent ecological conservation efforts in China. This policy is noteworthy because unlike earlier "one-dimensional" development policies, "ecological migration" reflects a certain amount of concern for ecosystems. However, based upon the premise of rehabilitating damaged ecosystems or preventing ecological degradation, the policy aims to restrict or transform, and in some cases to stop, traditional subsistence patterns and lifestyles in affected areas by persuading the inhabitants to migrate to other areas. This is the essence of "ecological migration".

A variety of problems have arisen due to the implementation of the "ecological migration" policy. Even if this policy were successful in promoting national integration, by forcing the diversity of subsistence patterns and lifestyles in a single direction by "homogenizing" them, it is doubtful whether the original goal of ecological conservation can be achieved. In order to resolve this doubt, it is therefore crucial to gain a comprehensive understanding of the actual state of ecological migration.

A nation of migrants and its remote regions

China, a nation of migrants

The Chinese term for "ecological migration" *(shengtai yimin)* literally refers to both the practice of migration conducted for the purpose of conserving the ecology and the people (migrants) who are subject to the various activities that accompany migration. So what does "ecology" *(shengtai)* mean in this context? Like many other modern Chinese words, it has been borrowed from Japanese, or more correctly, it has been re-imported into Chinese from Japanese *(seitai)*. As opposed to "ecology", the word "migration" is a familiar one in both Chinese and Japanese. When speaking of Chinese migration *(yimin)*, overseas Chinese is likely to be the first connotation. However, in Chinese the term for "migration" is not necessarily restricted to an "overseas" context.

A notable academic text in the field of internal Chinese migration is, "History of Chinese Migration", which consists of six volumes (Ge 1997a, 1997b; Wu 1997a, 1997b; Cao 1997a, 1997b). In the book, the term "migration" was first used in Zhouli's, "The Rites of the Zhou Dynasty", which is considered to have been written in the late Zhanguo period more than 2,000 years ago. At that time, the term was not used as a noun, but as a verb in the context of compelling people affected by a food crisis in their homeland to relocate to places where food was more abundant as a special relief measure (Ge 1997a: 3). From the rise of the Qin Dynasty, the first centralized administration in China, down to the republic of today, migration has occurred in various forms within China. The migrations that have shaped the China of today have all been driven by a range of factors including the acquisition of territory, the distribution of ethnic groups, and the characteristics of the ecological environment. In this sense, China can truly be described as a "nation of migration".

The roots of China's proud millennia-long history lie in the valley of the Huanghe (Yellow River). Since the climatic conditions in the Huanghe Valley were more conducive to human subsistence and prosperity than the hot, humid Changjiang Valley or the country's cold, dry northern region, agriculture flourished there sustaining a large population around the first century BCE. As the population grew, landless peasants began to appear. For these people, the warm southern regions were more appealing than the cold north of the country. During the approximately 1,600-year period from the rise of the Qin Dynasty in 211 BCE to the collapse of the Yuan Dynasty in 1368, the main flow of migration was from the Huanghe Valley to the Changjiang Valley. Over this time, the ratio of the population in the south to that to the north (where the border between the north and south are defined by the Huai River and Mt. Qinling) rose dramatically; from 1:4 in the Early Period (early Han Dynasty) to 4:1 in the Later Period (Yuan Dynasty). Between the founding of the Ming Dynasty in 1368 until approximately 1850, not only the Huanghe Valley, but also almost all of the southern plains, including the Changjiang Valley, became highly populated and the traditional form of plains agriculture was unable to support the increasing population. As a result, peasants began to leave the plains for the mountainous areas where they cleared new land for cultivation. This movement was accelerated in the 16th century by the introduction from the Americas of new crops such as sweet potatoes, corn, peanuts and potatoes that were suited to cultivation in mountainous areas. Associated with the increase in migration to the mountains from the early 18th century, the vast areas of virgin forest from Mt. Qinling to Mt. Nanling, between the Changjiang and the Zhujiang Valleys and from the Zhemin Hills to the Yungui Plateau, were harvested, the natural vegetation was removed, and almost all arable land was planted with corn and sweet potatoes. This attracted ever-greater numbers of migrants to the mountains in search of food or wealth and consequently, even in these mountainous areas, the population reached levels approaching the carrying capacity of the environment. Thus, in the period between the Taiping Rebellion in 1851 and 1950, large numbers of migrants migrated to the three northeastern provinces, Inner Mongolia, the northwestern provinces and Taiwan (Ge 1997a: 43–47, 66–67).

Remote regions as migrant destinations

These destinations of Chinese internal migration *(Han migration)* since the Ming and Qing Eras are essentially geographically equivalent to the migration events to the remote regions of modern China. This trend continued through the period of Kuomintang rule and has continued in the present era of the People's Republic of China (PRC). As a result of the influx of large numbers of Han migrants, these regions were gradually converted to agricultural use and "Hanized", and were subsequently transformed into territories that were inseparable from the rest of the nation. However, the practice of agriculture in areas unsuited to cultivation has lead to serious environmental problems such as soil degradation, desertification of grasslands, and the drying up of rivers.

The majority of advances in agriculture began in the Ming era, when Han migrants in Guizhou Province on the Yungui Plateau clashed violently with the indigenous ethnic peoples. In the 270-year Qing Period alone, the population of Han Chinese in Guizhou Province increased more than tenfold (Luo 1993). The influx of large numbers of Han Chinese migrants to Yunnan also commenced in the Ming era (Xie 1996: 24), and even into the Qing Era, migration from Sichuan, Guizhou, Hunan and Quangdong Provinces showed no signs of abating. Between 1661 and 1749 the population of Yunnan increased by a factor of five (Cang 1998).

It was from the early Qing Era that Han migration to the region north of the Great Wall of China was at its highest. In the first half of the Qing Period, the Qing Dynasty officially prohibited Han migration to the homelands of the Manchurians (in the northeast) and the Mongolians, with whom the Manchurians had made alliances. However, in reality, migration from the interior still occurred. For example, the people who relocated to the northeast consisted primarily of migrants leaving the Shandong and Hebei Provinces in search of employment, and prisoners from all over the country. There was also migration of Han Chinese into southern Inner Mongolia, from Shanxi Province to the area of Tümed *(Tumote)*, and from Hebei Province to the area of Chahar *(Chahaer)*. As a form of recognition that Han migration had occurred, the Qing Dynasty replaced the banner system with a new local administrative structure for migrants within Inner Mongolia that was divided into prefecture *(Fu)*, sub prefecture *(Ting)*, department *(Zhou)* and county *(Xian)* levels, indicating that the settlement of Han migrants was encouraged. In the mid-18th century, after defeating the Oirad Mongols, the Qing Dynasty actively promoted migration to the northwestern region, in particular to Xinjiang. The Qing Dynasty exerted pressure on large numbers of peasants, mainly from Gansu, Shanxi and Sichuan Provinces, to migrate to Xinjiang. At the same time, the Qing changed its policy on prisoners by sending them to the northwest rather than to the northeast, which had been the destination of prisoners up until that time. Incentives to promote migration to Xinjiang were employed. For example, in addition to having their sentences reduced, prisoners serving their term in Xinjiang were permitted to be accompanied by their families and were assured that they would be considered *mintun*, as having the status of commoner, after their release As a consequence, more than half the population of Xinjiang in 1777 consisted of Han migrants and their descendants (Cao 1997b: 472–489; 493–495).

In the latter half of the Qing period, the law prohibiting migration to the northeast region was abolished and the number of migrants to that region increased suddenly as a result. Furthermore, due to the inequity of the Boxer Protocol Treaty, the Qing Dynasty was obliged to pay reparations amounting to 450 million *tael* to the foreign powers. In order to assist with paying its share of the reparations, Shanxi Province sought permission from the Qing rulers to cultivate the grassland areas of Inner Mongolia, in Ulaanchab *(Wulanchabu)* and Ih Juu *(Yikezhao)* Leagues, and Chahar Banner. On receiving the official approval of the Qing rulers, Shanxi Province began to cultivate almost all of the grassland areas of western Inner Mongolia. Furthermore, according to records of the early Republic of China, referring to the then Outer Mongolia (now officially Mongolia), "the number of Han migrants has exceeded 100,000, of which 50,000 are agricultural migrants". The Mongol population for this area was 540,000 (Cao 1997b: 505).

Even after the formation of the Republic of China, land-clearing projects in the outer regions did not abate. Rather, they were conducted even more aggressively. In the late 1920s, the Kuomintang Government created three provinces in Inner Mongolia: Rehe, Chahar and Suiyuan. In addition to migrants from Shanxi and Hebei Provinces, many people came from Shandong and Henan Provinces. Thus, increasingly large areas of grasslands were cultivated. In 1931, in Suiyuan Province, "More than 180,000 *chin* (1 *chin* = 6.667 hectares) of wasteland was released… and in the 1930s the population of the province grew to 2.033 million". Also, in Chahar Province, due to the rapid increase in the number of Han Chinese migrants from Hebei Province and other places, "As of 1927, wasteland amounting to 66,000 *chin* have been put under cultivation, cultivated lands are connected together, villages are densely concentrated, urban development is progressing on an unprecedented scale, and more than half the total area of the province is settled by migrants" (Cao 1997b: 509–510). In the era of the Republic, Xinjiang also received a large number of Han migrants who had been affected by natural disasters in Gansu, Shanxi and Henan Provinces.

During the latter half of the Qing Dynasty the influx of migrants to the remote regions continued, and many migrants within the remote regions moved from the plains to the mountainous areas. Initially, migrants would colonize the plains where the natural conditions were more favorable, but as the population levels approached saturation, the migrants tended to advance into the mountainous areas. This process of advancement was basically autonomous. Since "the majority of migrants were poor peasants, landless or illiterate, with the rest made up of unemployed idlers", they did not do even the minimum required preparation or investment for migration (Ge 1997a: 68–69). The way in which these migrants, who were seeking immediate success or gain, cleared land for cultivation was extremely *ad hoc* and exploitative of resources; "in the process of their land clearing, natural and land resources and the ecological environment were devastated" (Ge 1997a: 68–69). For example, before the area of Weichang (now in Weichang Manchu-Mongolian Autonomous County, in Heibei Province) on the southern Mongolian Plateau was cultivated, it abounded with trees, rivers and wild animals, and served as the hunting grounds of the imperial family. Soon after cultivation began at the end of the Qing period, the population increased rapidly. In 10 years, the total population grew by 82 percent, reaching 89,000 in 1917. By 1934, the area of land under cultivation had expanded to 1 million *mu* (1 *mu* = 0.67 hectares) and the natural environment was completely transformed. After large numbers of trees were harvested and vast areas of land were cleared for cultivation, desertification began, becoming increasingly severe with each passing day (Cao 1997b: 20; 503).

In the era of the PRC, the migration of Han Chinese to the remote regions was conducted in a planned and systematic manner to meet national construction needs and military requirements. Soon after the formation of the Republic, Mao Tse-Tung and the Communist Party Central Committee launched the slogan "Open up and protect the remote regions". Retired soldiers of the People's Liberation Army and young people from cities and villages throughout the country were relocated to remote regions such as Xinjiang, Inner Mongolia, Heilongjiang, Yunnan and Guangxi, to help construct over 2,000 state-run farms (Liu Bingfeng 2004). Particularly in Xinjiang, the state assumed control of the long-standing Tuntian system in which garrisoned troops or peasants opened up wasteland, grew cereals and organized an extensive, permanent land reclamation collective, known as the "Xinjiang Production and Construction Corps". It was reported that at the end of 1999, well after this movement had peaked, there were still 2.42 million people in the Xinjiang Production and Construction Corps (Gao 2001). In addition, refugees from all over China also moved to Inner Mongolia and Xinjiang due to natural disasters and other reasons, "increasing the population by a factor of four to five in 50 years: in Inner Mongolia from 4 million in 1930 to 12.39 million in 1964 and 19.27 million in 1982; in Xinjiang from 2.5 million in the 1930s to 7.27 million in 1964 and 13.08 million in 1982" (Huang 1987: 73). Even if the periods, routes, motivations of these migrations differed, the migrants invariably took up agriculture in the areas that they settled.

The population pressure and the expansion of agricultural area brought about by these migrations not only caused considerable changes in the subsistence patterns and lifestyles of the indigenous ethnic inhabitants, but also disturbed the balanced relationship that had long prevailed between humans and nature. Although different to the earlier reckless reclamation of the Han migrants, the basic policy promoted by the central government in the remote regions also favored agriculture and discriminated against livestock herding, which meant that the mentality of the migrants and the government had much in common. Some scholars, such as Liu Xuemin of the College of Resources Science & Technology, Beijing Normal University, contend that the current degraded state of the ecological environment in western China is a consequence of the national policies of the past. The following is the analysis of Liu Xuemin.

> Starting in the 1950s, China implemented large-scale reclamation of grasslands on three occasions in which pastures were cleared to cultivate cereals. On the first occasion in the 1950s, during the period of the People's Commune, extensive grassland clearance was undertaken in an attempt to establish agriculture on a large scale. As a consequence, winter and spring grazing lands were reduced, and the soil suffered progressive desertification. On the second occasion in the 1960s and 1970s, during the period of the Great Cultural Revolution, grasslands were again recklessly cleared, on the basis of misguided slogans: "Herders must cultivate their own cereals" *(mumin buchi quixinliang)* and "Grazing areas must evolve into agricultural areas" *(muqu xiang nongqu guodu)*. The result was further deterioration of the ecological environment. Finally, in the years since the reform and opening up policy, large-scale cultivation was again undertaken in many grassland areas as part of local initiatives directed at producing short-term gain. Since this policy was implemented, initiatives such as the "Food Self-sufficiency Project *(midaizi gongcheng)*" and "Vegetable Self-sufficiency Project *(cailanzi gongcheng)*" were promoted, even in areas that were fundamentally unsuitable for the cultivation of agricultural crops and vegetables. As a result, the ecosystems were degraded further (Liu Xuemin 2002: 47).

Furthermore, in the early 1980s, soon after the introduction of the reform and opening up policy, a contract system that was being promoted in the eastern agricultural areas was rapidly adopted in the Inner Mongolia Autonomous Region. As part of this system, it was decided that domestic animals and grassland grazing rights would be distributed to individuals. Consequently, in grasslands that were originally undivided, a new method of grassland utilization (known as *caokulun*) was established in which the grazing land of each household was artificially divided by barbed wire fencing. In a short space of time this method was adopted in livestock raising areas throughout the country. The concept of grazing animals in a fixed place is referred to as "settlement farming". As a result of this style of grazing becoming accepted practice, the traditional grazing style in which herders moved from place to place to rest grazing lands (nomadic herding or nomadic grazing) was effectively eliminated. As a consequence, within 10 years desertification was clearly evident in various parts of Inner Mongolia and eventually storms of "yellow sand" carried the sand from these areas and deposited their loads as far away as the inland areas of eastern China, creating a serious environmental problem. In this way, a chain reaction was initiated, beginning with land clearance by immigrants, which in turn led to the shrinking of the grazing lands and the adoption of settlement grazing with the concomitant discontinuation of the practice of nomadic herding. These events culminated in the desertification of grasslands and the degradation of ecosystems and resulted in the phenomenon of yellow sand. It can therefore be concluded that agricultural-style grazing, which was adopted in preference to traditional livestock raising, was one of the leading factors underlying the destruction of the ecological environment in the remote regions of China.

Environmental policy and ecological migration

Background to the emergence of ecological migration

It was the indigenous ethnic inhabitants of the remote regions, those traditionally engaged in raising livestock that bore the brunt of the negative legacy of this change in agricultural practice. Under the pretext of restoring and conserving degraded ecological environments, the Chinese government launched a campaign of "ecological migration", under which indigenous inhabitants were pressured to leave what little land they had left, and migrate elsewhere.

"Ecological migration" has been effectively practiced since the 1980s, long before the large-scale campaign was initiated. In one district in the southern mountains of the Ningxia Hui Autonomous Region, designated by the state as a "specially poor district", severe environmental degradation made it difficult for the inhabitants to subsist. Since 1982, under the guidance of the national government, the inhabitants were encouraged to migrate from the region. This marked the beginning of the practice of "ecological migration" in China. The measure, adopted at Ningxia, was introduced in other "specially poor districts" from 1986. The efficacy and necessity of the policy of "ecological migration", which was originally aimed at the eradication of poverty, gradually became widely accepted by the people (Li Ning et al. 2003).

From the end of 2004 more than 50 scientific papers on the subject of "ecological migration" have been published, and many academic dissertations have also taken up "ecological migration" as a theme (Meng Linlin 2004; Hu Huazheng 2004). The term "ecological migration" was coined in a scientific paper in 1993 within the context of the migration associated Sanxia Dam. Although the paper does not necessarily define the term clearly, the concept of "ecologic migration" was proposed as a solution to the problems of a deteriorating ecological environment and increasing population pressure (Ren et al. 1993).

In the early years of its promotion, in addition to ecological conservation, the objectives of "ecological migration" included poverty eradication and the promotion of dam construction. It has only been since 2000 that the main focus of the policy has been that of ecological conservation and numerous reports on the policy have been presented to the public through newspaper articles under the title "ecological migration". Two examples are the migration from the Shennongjia Nature Reserve, established for the protection of rare wildlife (Zheng 2000), and the migration of herders from Alasha League in the Inner Mongolia Autonomous Region (Wulantuya 2000).

This was the exactly the same time that the West Development Project was initiated. One motivation for this project was the need to address the degradation of ecosystems. According to Du Ping, three factors directly contributed to the establishment of the West Development Project, namely the Asian financial crisis of 1997, the massive flooding of the Changjiang Valley in 1998, which resulted in more than 200 billion yuan of direct damage, and the storms of yellow sand in 1999 that caused severe damage over large areas of inland China (Du et al. 2004: 17).

In short, these ecological environmental problems inevitably became one of the main concerns of the West Development Project. The five pillars of the West Development Project policy mentioned in China's 10th five-year plan (2001–2005), released in March 2001, were: (1) Acceleration of infrastructure construction; (2) Improvement and maintenance of ecological environments; (3) Coordination and rationalization of industrial structure; (4) Development of scientific technology and education; (5) Deepening of reforms and further opening up to the outside world. Of these priorities, the central government is stressing the importance of (1) and (2) and has assigned them as tasks deserving the highest priority (Onishi 2004: 48).

Ecological migration as policy

As described above, the central government was directly responsible for the ecological environmental problems. On December 14, 2002, Premier Zhu Rongji officially announced the "People's Republic of China State Council Decree (No. 367)". This law incorporated the "Cultivation discontinuation for forest (grassland) restoration ordinance" (this ordinance is designed to discontinue cultivation of affected land, in order to reforest it or develop it as grassland) *(tuigeng huanlin(cao))*. This ordinance explicitly mentions "ecological migration" in a number of places. For example, in Article 4 it states that, "cultivation discontinuation for forest restoration measures shall place priority on ecology... its implementation must be combined with ecological migration". Article 54 states that,

> in the process of implementing cultivation discontinuation for forest restoration, the government shall encourage ecological migration, and provide assistance for the livelihood and productivity of farming households that participate in ecological migration (Zhonghua Renmin Gongheguo Guowuyuan 2002).

Following on from the "cultivation discontinuation for forest restoration" initiative for agricultural areas, a similar initiative – "grazing discontinuation for grassland restoration" *(tuimu huancao)* – was launched in 2003, which resulted in a stream of ecological migrants being relocated to livestock raising areas. "Grazing discontinuation for grassland restoration" means to relinquish the grazing of livestock in order to allow the grazing lands to be restored to natural grassland. According to a report by Wang Daming, this term was proposed as early as 2001 in the context of ecological conservation in livestock raising areas of Qinghai Province. A related proposal aiming to discontinue low-cost, low-efficiency, traditional livestock raising and to construct so-called "man-made grasslands," through the substitution of inferior types of natural grazing with high-quality natural grazing in order to enhance the energy content and nutritional value of the fodder (Wang Daming et al. 2001). Subsequently, based on the experience of the "cultivation discontinuation for forest restoration" initiative, and also by accepting the proposal of the Chinese People's Political Consultative Conference that "grazing discontinuation for grassland restoration" should be implemented in livestock raising areas (Liu Zenglin 2002), the central government decided to extend this initiative to other areas, including eastern and northern Inner Mongolia, northern Xinjiang and eastern areas of the Qinghai-Tibetan Plateau, and also to supply livestock feed (National Development and Reform Commission et al. 2003). This development encouraged debate on proceeding with the creation of "man-made grasslands" (Abulizi 2003; Chen 2004).

The measures implemented under the name of "grazing discontinuation for grassland restoration" can basically be divided into three categories. These are the "banning of grazing" (completely prohibiting grazing for a fixed period of time), "suspension of grazing" (discontinuing grazing from the time of grass germination to the time of seed maturation) and "rotational grazing" (dividing grazing lands into several partitions, according to natural features and human judgment and to allow grazing by rotation in each partition) (Wang Xiangyang et al. 2003). The "grazing discontinuation for grassland restoration" project set a target of rehabilitating 1 billion *mu* of land, an area equivalent to 40 percent of all the degraded grazing land in China, within five years of its official commencement in 2003 ("Zhongguo Muye Tongxun" 2003). Thus, with the sharp rise in the number of "ecological migrants" generated by policies aiming at the conservation of ecological environments, such as "cultivation discontinuation for forest (grassland) restoration" and "grazing discontinuation for grassland restoration", "ecological migration" is attracting an increasing amount of attention.

Diversity of ecological migration

In considering ecological migration, it is not necessary to limit considerations of its objectives and efficacy to the conservation and restoration of the ecological environment – the aspects that have received close scrutiny since 2000.

This is because ecological migration is intertwined with a wide range of "non-ecological" elements. For example, numerous local government bodies implement "ecological migration" exclusively in the context of poverty eradication (Dongribu 2000). Currently, other than in the previously mentioned "cultivation discontinuation for forest restoration ordinance," the word "ecological migration" rarely appears in related laws, such as the "Environment Protection Law," the "Forest Law," the "Sand Prevention Law" and the "Grasslands Law" (Standing Committee of the National People's Congress 1989, 1998, 2001, 2002). Furthermore, the term "ecological migration" has never been clearly defined by the government.

In light of this situation, researchers have made various proposals about how "ecological migration" should be defined and classified. For example, there is a popular trend to interpret "ecological migration" as a sociological phenomenon (Sang 2004; Wang Peixian 2000: 26). On the other hand, there are claims that the individual subjects of this sociological phenomenon – that is, the affected farmers and herders – should also be incorporated in the definition of ecological migration (Meng et al. 2004). Some scholars point out that the essence of ecological migration should be understood from the standpoints of both efficient cause and final cause. Here, efficient cause refers to the degeneration of the natural environment and excessive population growth, while final cause refers to conservation of ecological environments and improvement in the livelihood and productivity of all herders (Li Xiaochun et al. 2004: 35). It has been suggested that "ecological migration" can be classified into five categories on the basis of purpose: to protect the catchment areas of large rivers; to prevent sand storms; to prevent water disasters; to construct hydropower facilities; to solve poverty problems; and to protect rare wildlife and tourist sites (Pi 2004: 58–59).

While defining and classifying "ecological migration" may assist in the understanding of this phenomenon, at this point, it is perhaps more helpful to consider how "ecological migration" actually occurs in specific places and times rather than to extract common aspects from a range of cases. This makes it possible to analyze the conceptual background of "ecological migration" from a variety of angles instead of adopting a one-dimensional, rigid approach to the subject. In addition, this facilitates dynamic investigations into the effects of this phenomenon on local communities and the nation as a whole.

Western remote regions and ecological migration

The "heterogeneous" remote regions of the west

The areas targeted by the West Development Project launched in 2000, are principally in the geographical west of China: Chong Qing City, Sichuan, Guizhou and Yunnan Provinces, and the Tibet Autonomous Region (all in the southwest); Shanxi, Gansu and Qinghai Provinces, the Ningixia Hui Autonomous Region, and the Xinjiang Uighur Autonomous Region (all in the northwest). In addition to these 10 regions, the Inner Mongolia Autonomous Region and the Guangxi Zhuang Autonomous Region were also included, making a total of 12 regional self-governing bodies. Apart from all these, the Enxi Tujia and Miao Autonomous Prefectures in Hubei Province, the Xiangxi Tujia and Miao Autonomous Prefecture in Hunan Province, and the Yanbian Korean Autonomous Prefecture in Jilin Province, located in the far northeast of China, near the border with North Korea, also come under the umbrella of the West Development Project.

As is indicated in the list above, the term "west" in "West Development Project" does not literally mean the geographic region in the west of China. Rather, it refers to a number of features that distinguish the "west" from eastern China. That is, it refers to regions that are not politically centralized, have economies that are not agricultural, with cultures that do not use Chinese writing or language (*hanwen* or *hanyu*), or with inhabitants whose ethnicity is not Han. Therefore, since the east is considered to be the center of China, the "west" is regarded as "remote" and "heterogeneous".

From an exotic perspective, it might be desirable that the heterogeneity of the west is preserved. However, in terms of a nation, there are good reasons for not encouraging this heterogeneity. While the population of the heterogeneous western remote regions amounts to only 27.4 percent of the national population, the west occupies 71.4 percent of China's total land area. Most of China's ethnic minorities and energy resources are dispersed in the west. Furthermore, the west is viewed as the source of disasters, as typified by the yellow sand storms that have plagued the east; a commonly held belief is that such disasters are the result of the destruction of ecosystems brought about by the heterogeneous subsistence patterns and lifestyles of the inhabitants of the western regions, particularly the ethnic minorities.

Thus, in order to protect the natural environment from destruction by these heterogeneous inhabitants, it is important to exercise control over them first. Many researchers perceive that ecological migration is the very basis for rehabilitating the ecosystems of the west. For example, one researcher claims that environmental problems exist because the subsistence patterns and lifestyles of the ethnic minority inhabitants of the "west" are backward *(luohou)*, and that these people should therefore be reformed. The argument for this is as follows:

> The vast majority of the western regions of our country, in which the ecological environment is degenerating, are areas of poverty and are inhabited by ethnic minorities. Their backward subsistence patterns and lifestyles, which they have used for a long period of time is one of the major causes of environmental degradation in the affected areas. To protect the ecological environment it is first necessary to reflect on the subsistence patterns and lifestyles of the people in these areas and then reform them… (Chi 2004: 14–15).

Many researchers share the same view; heterogeneous characteristics and ways are not merely different, but "backward". For example, one historian stresses the advanced state of agriculture of the Han Chinese and the necessity of promoting agricultural development and Han migration as contributing toward national integration as follows:

> In the era before modern industry was developed, agriculture was undoubtedly the most advanced and most reliable industry. Agriculture is highly significant in the flourishing of an ethnic population and their economic and cultural progress. For the Chinese nation *(zhonghua minzu)*, agriculture is a common material foundation, and at the same time, the source of the tremendous cohesive force of the Han people (Ge 1997a: 96).

These researchers who consider ethnic minorities heterogeneous and "backward" and Han Chinese as "universal" and "advanced" share a common belief in social evolutionism. This philosophy would claim that in regard to subsistence patterns, agriculture is more advanced than raising livestock; in living patterns, a fixed residence is more advanced than a mobile tent; and in living environments, urban is more advanced than rural. It also assumes that less developed areas (west) inevitably evolve towards being more advanced (east). This idea has been raised by many researchers, including Chinese scholars based overseas. As an example, a paper by Wang

Ke contains the following phrases: "Rich east and poor west", "Advanced east and backward west" and "The supporting east and the supported west". This way of thinking is clearly present in the West Development Project too, and it is accepted as axiomatic that the east is the homeland of the Han Chinese. In the Law of the People's Republic of China on Regional National Autonomy, revised in 2001, the subject of autonomy was modified from "ethnic group" *(minzu)* to "area" *(diqu)* i.e. an ethnic autonomous region. This revision clearly exposes the Chinese government's belief that ethnic minorities should study Chinese and become assimilated with the mainstream culture of the country. Through economic integration, the West Development Project is undoubtedly implementing a form of national integration that is aiming to change the cultural and ethnic consciousness of the western ethnic minorities (Wang Ke 2001: 57–58).

It is in the promotion of "national integration" that the ecological migration measures of the West Development Project are most effective. This is because, unlike the various slogans previously used to promote national integration, the term "ecological migration" has the ability to overpower all doubts and opposition – the connotations of common human welfare carried by the word "ecology" justifies this belief. People who lived by herding livestock are pressured to move as ecological migrants to small towns *(Xiaochengzhen)* constructed within their own areas or elsewhere. There, they are made to live in densely populated residential areas and to work in industry or in service jobs or other tertiary sector employment. This change in lifestyles and subsistence patterns reduces the dependency of people on the ecological environment and thus to promoting ecological conservation (Wang Peixian 2000; Liu Xuemin et al. 2002). This is the scenario painted by the ecological migration policy. Ecological conservation is realized through the complete eradication of the "heterogeneous presence". The process of "ecological migration" can be seen as a series of processes directed at the elimination of this "heterogeneity". Whether or not it actually results in ecological conservation is uncertain and the effectiveness of "ecological migration" as an ecological instrument awaits further verification.

The "homogenizing" effect of ecological migration

Almost all the researchers currently studying "ecological migration" in China today tend to approve of the policy, advocate its necessity, and consider how it can be successfully implemented. While there is virtually unanimous approval for the policy overall, there are some differences in opinion regarding the details of the policy. These views can be broadly categorized into two schools, the "enthusiastic school" and "cautious school".

The "enthusiastic school" considers that the active implementation of migration itself is the key to successfully solving environmental problems. There are, however, some differences of opinion about whether displaced people should migrate to other places within their region, or to other regions. Some argue that migration should be limited to the region originally inhabited by the affected people, that is, in the vicinity of the "migration origin". In this view, it is also necessary to promote urbanization and changes in the structure of industry by constructing small towns to accommodate the displaced migrants that have the five major infrastructural categories: roads, water supply, telecommunications, electricity and radio and television. Others argue that the migration destination should not be limited to neighboring areas of the migration origin, but rather that migration should be from "west" to "east". This view proposes the construction of new migrant villages and towns in two areas in good environmental condition: the northeast plains, and the plains along the mid to lower reaches of the Changjiang River. Migrants from the Loess Plateau would be relocated to the former area, while migrants from the Yungui Plateau would be relocated to the latter. The reasoning behind this is that since the "east" enjoys the effective benefits of environmental conservation measures such as "cultivation discontinuation for forest (grassland) restoration" it has a duty to accept ecological migrants from the "west" (Fang 2001: 40; Fan et al. 2003: 36). Accordingly, it has been proposed that the national government draft an "ecological migration law" and clearly legislate the east's obligation to accept ecological migrants (Xu Suhuan 2003).

In contrast, the argument of the "cautious school" is based on the experience of Alasha League in the Inner Mongolian Autonomous Region and maintains that the positive effects of "ecological migration" projects are limited to the migration origin, and even then only for the short-term and to a limited extent. At the same time, there also negative effects, such as new environmental problems like soil alkalization, caused by excessive groundwater consumption in migrant destination areas and a sharp increase in birth rates. Thus, the basic position of the "cautious school" is that, since ecological migration may have a net negative effect on the ecosystem as a whole, it should be used not as a strategic policy, but only as a supplementary measure (Xu Honggang 2001: 25).

The "enthusiastic school" and the "cautious school" have one element in common; both have restricted their analyses to the technical aspects of the phenomenon of "ecological migration". On that basis alone, they then evaluate the present situation and try to forecast future developments. They unilaterally try to determine how "ecological migration" should best be implemented and have little or no regard for the people that will be subjected to the policy. They believe that the reason why many ethnic minority inhabitants of the west have become ecological migrants – leaving their living environments and being forced to abandon their traditional culture – is that their culture, as represented by their subsistence patterns and lifestyles, is backward. This conclusion is supported by the fact that, while the numerous slogans advocating respect for the traditional culture of the ethnic minority groups, have not yet completely disappeared, they seem to have been replaced by the new buzzword "ecological migration" and are now almost never heard.

Perhaps being conscious of this situation, a few ethnic minority researchers have raised the issue of the rights and interests of the ethnic minorities within the context of "ecological migration". One such right is the right to preserve their traditional culture, which is guaranteed by law (Wuligeng 2003). Some researchers have gone a step further, by pointing out that positioning the "ecological migration" policy within the framework of regional economics is not scientifically rational, and also that the lack of reflection on the policy by the regional leaders that are implementing it has caused a reduction in the income of livestock herders. In this way, these scholars are sounding an alarm and are warning that such economic problems, if left unaddressed, may develop into ethnic problems (Gegengaowa et al. 2003: 120).

In the context of the relationship between ecological migration and ethnic culture, and based on the experience of the Inner Mongolia Autonomous Region, Gegengaowa et al. (2003) claim that the idea that raising livestock and that the cultures of ethnic minorities are equated with backwardness and should be altered or abandoned is a dangerous one. At the same time, they raise questions regarding the ethical responsibility of researchers, as the quote below shows.

> The community of theoretical researchers in Inner Mongolia has an honorable tradition of defending the legitimate policies of parties and governments. Due to this, the Inner Mongolia Autonomous Region has been historically recognized as a model autonomous region. Still today, it is a model of national unity / amity between nationalities *(minzu tuanjie)* and social stability. However, these researchers, who are viewed as "yes men" by the herders, do not think about issues from the standpoint of the herders. In fact, they pander so much to authorities that their words risk steering government policy relating to livestock raising areas in the wrong direction (Gegengaowa et al. 2003: 120–121).

Certainly, the Inner Mongolia Autonomous Region, which was created in 1947, prior to the establishment of the PRC, has been regarded as a model of national integration in the process of China's efforts to unify various nationalities (or ethnic groups) into a single nation. Similarly, in the current process of national integration, Inner Mongolia is expected to serve as a model of unification in ethnic minority areas, and in Inner Mongolia Autonomous Region government is attempting to fulfill this role. For example, by 2000 the Inner Mongolia Autonomous Region had already implemented the "grazing discontinuation for grassland restoration" project in Alasha League in the west of the region. At this time, the regional government pressured more than 2,000 herders, accompanied by 150,000 head of livestock, to leave the Helan Mountains as ecological migrants (Liu Jun 2000). Also, within the entire region of Inner Mongolia, the government has actively implemented "cultivation discontinuation for forest (grassland) restoration" and "grazing discontinuation for grassland restoration". The "grazing discontinuation for grassland restoration" project, officially launched in 2003, was implemented in 65 banners and counties under the jurisdiction of 12 leagues and cities – practically the entire area of Inner Mongolia (Xu Feng 2003). The project was actively pursued in 33 livestock raising banners and in 21 half-agricultural, half-livestock raising banners and counties. The total scale of "grazing discontinuation for grassland restoration" in these areas will amount to 600 million *mu*. In the first implementation period, from 2002 to 2010, 450,000 *mu* is expected to be restored, with a further 150,000 *mu* restored in the second implementation period, from 2011 to 2015 ("Dongwu Kezue yu Dongwu Yixue" 2003). Furthermore, the government of the Inner Mongolia Autonomous Region announced a plan to implement ecological migration of 650,000 people within a six-year period from 2002 (Yin Yue et al. 2002).

There has been growing concern in China in recent years about environmental degradation in the Inner Mongolian grasslands, which are the closest grassland areas to the heart of eastern China – the areas around Beijing, Tianjin and other areas in the North China Plain. Consequently, Chinese society as a whole has been paying close attention to the issue of ecological migration in Inner Mongolia. As of the end of 2004, the term "ecological migration" appeared in the title of more than 50 newspaper articles, with more than 20 percent of these articles concerned with "ecological migration" measures in the Inner Mongolia Autonomous Region. Addressed to the people of the "east", as if to ease their anxieties, these articles emphasize how the "backward" livestock raising methods practiced in the west had been abandoned and how herders had been resettled in cities, describing these measures as "success stories". The reduction and disappearance of the heterogeneous practice of livestock raising and herders from Inner Mongolia signifies conformity to the standards of the "east", or more accurately, "homogenization". In this sense, Inner Mongolia could be seen as a new model for the process of homogenization of the heterogeneous "west".

The potential of "heterogeneous" practices

Conversely, there is a view that even if the storms of yellow sand that plagued Beijing and other areas on the North China Plain do originate in Inner Mongolia, this natural phenomenon is not produced by nomadic culture, but rather by the cruel elimination of agrarian culture. Based on historical records of migration into Inner Mongolia, the organization Hanhaisha (an NGO that works to protect ecological environments and ethnic traditional culture in desertified areas of China) argues that the peasants who migrated into the region were not capable of clearing land properly. As their population expanded, the space available to nomadic herders came under pressure, eventually leaving them unable to sustain a livelihood. This is the underlying reason for the growing severity of the yellow sand storms. In this sense, Hanization equals desertification, and settlement culture equals disaster. Nomadic herding, in fact, can reduce the stress on grasslands, and contribute to the conservation of ecological systems. In the nomadic culture of traditional herding, the most precious resources are water and grass, followed by livestock. This ecoculture harmonizes humans and nature as one and thus represents an elevated state to which human beings should aspire (Hanhaisha 2004).

When we think about environmental problems such as yellow sand storms, we almost never consider them as being related to human culture, but rather as if they were completely separate and independent phenomena. However, as Hanhaisha claims, this perception does not reflect reality – the yellow sand storms may be the penalty incurred for the extinction of nomadic herding culture. The essence of environmental problems is cultural; environment and culture are really two sides of the same coin.

As I have already briefly mentioned in the first half of this chapter, migrants who relocated from inland China to remote regions cleared land for cultivation and, in doing so, damaged the entire ecological environment in the areas in which they settled; a fact that has been acknowledged by many researchers If this is the case, efforts should have been made to conserve the "ecology" of the region by reviving the traditional subsistence patterns of the indigenous people who had become displaced. However, the current "ecological migration" measures that aim to protect the "ecological environment" have the opposite effect. Consequently, when thinking about "ecological migration" as an ecological issue, it is important to consider who is making an issue of "ecology", and on what grounds. Most importantly, there is no such thing as a completely independent "ecology" that can be protected, one that is free from all human interference. "Ecology" includes people's views of nature and their subsistence patterns. In this sense, rather than being an objective matter, "ecology" is an extremely subjective and cultural issue.

Cultural diversity is an integral part of the richness of human experience, and it is also a pivotal factor in the relationship of humans to nature. If a people of culture "A" fail to maintain a harmonious relationship with nature, the example of a people from culture "B" who have maintained harmony may help the former people reconnect with nature. Having options, such as A and B, is essential to maintaining a constant harmony with nature in the face of change. Respect for the various ways in that people have developed different cultures under different natural conditions enhances prospect of the continued survival of humanity. Given China's wealth of ethnic diversity, in this age of environmental conservation it will be expected to increase this kind of cultural contribution to the world.

China's current leadership is currently publicizing a slogan, "People oriented" *(yiren weiben)*, reflecting its desire to pursue not only economic growth, but also the sustainable and balanced development of the entire society. The slogan means "people first" and "giving priority to people". It goes without saying that people are not merely abstract entities as they always exist under specific cultural conditions. If "people first" and "giving priority to people" means "giving importance to culture," then it should not be just one's own culture that is valued and the cultures of others must also be respected. In China, the slogan, "people first" is directed primarily at scientific development, with an emphasis also on the development of "harmony between humans and nature". Back in the era of slogans such as "political ideology first" and "economic development first", the diverse cultures of ethnic minorities and the diverse subsistence patterns that formed the foundations of those cultures were marginalized and objectified through the ideas of social evolutionism or rationalism. Now, in the era of "people first", the significance of this diversity should be reconfirmed as a model of "the harmony between humans and nature".

About the composition of this book

Over the next 10 years, there are expected to be 10 million ecological migrants in China (Pi 2004: 60). Although this is just one strategy for conserving the environment, if the scale and the features of the target areas are considered, there is no doubt that this initiative will have major effects on China's future. Research into ecological migration will be increasingly required in the years ahead. To build a solid foundation for such full-scale research, it is essential to commence by ascertaining the current situation.

This book, which aims to review the current reality, is composed of three parts. The areas discussed in this book are the Inner Mongolia Autonomous Region, Gansu Province, the Xinjiang Uighur Autonomous Region and Guizhou Province. The contents of all chapters are based on data acquired from field works.

Part 1, "Questioning ecological aspects: Can 'ecological migration' achieve environmental conservation?" mainly examines the "environmental" aspects of the policy. It poses the essential question of whether the policy, aimed at environmental conservation, really achieves this aim.

Part 2, "Questioning economic aspects: Can 'ecological migration' achieve poverty reduction?" mainly discusses arguments that are centered on "economic" aspects of the policy. By looking at how the income and spending of migrants changed with relocation, the economic effects of ecological migration measures are evaluated. Since ecological migration is also promoted as a means to achieve "poverty relief" and "poverty reduction," this part considers whether it really achieves these goals.

Part 3, "Questioning cultural aspects: What kind of transformation does 'ecological migration' effect?" changes the focus to the "cultural" aspects of "ecological migration" and poses wide-ranging questions regarding the policy.

As described above, this book clarifies various problems associated with the implementation of the policy of "ecological migration", in terms of the three aspects of environment, economics and culture. This book has been produced with the desire to contribute to making the noble objective of environmental conservation more achievable. It should be noted that ecological migration as discussed in this book deals only with contemporary environmental policy in China. For this reason, the term "ecological migration" has been enclosed in quotation marks in the main body of the text, especially in chapter titles and when it appears for the first time in each chapter. Also, when presenting readings of proper names in languages of ethnic minorities in China, the editors of the book shared a common recognition that, as far as possible, such names should be rendered directly as pronounced in the original language, without using Chinese characters. According to this convention, in this book, "Ejina" and "Yugu" are written as "Ejene" and "Yogor", respectively.

References

Abulizi Yusufu, 2003, Guanyu Xinjiang shengtai yimin de yiyi he xingshi de chubu tansuo, *Xinjiang Daxue Xuebao (Shehui Kexue Ban)*, Di san shi yi juan, Di san qi, pp. 32–35. {Chinese}
(Abulizi Yusufu, 2003, Preliminary analysis of the significance and forms of eco-migration in Xinjiang, *Journal of Xinjiang University (Social Science Edition)*, Vol. 31, No. 3, pp. 32–35.)

Cang Ming, 1998, Qingdai hanzu yimin ru Dian yuanyin kaolü, *Qingshi Yanjiu*, Di san qi, pp. 97–100. {Chinese}
(Cang Ming, 1998, On the cause for Han Chinese immigration into Yunnan Province in the Qing Dynasty, *Studies in Qing History*, The Third Period, pp. 97–100.)

Cao Shuji, 1997a, *Zhongguo Yiminshi- Di Wu Juan,* Fujian Renmin Chubanshe. {Chinese}
(Cao Shuji, 1997a, *History of Chinese Immigration – Vol.* 5, Fujian People's Press.)

Cao Shuji, 1997b, *Zhongguo Yiminshi- Di Liu Juan,* Fujian Renmin Chubanshe. {Chinese}
(Cao Shuji, 1997b, *History of Chinese Immigration – Vol.* 6, Fujian People's Press.)

Chen Xin, 2004, Shishi tuimuhuancao gongcheng, jiakuai rengongraodi jianshe shi guanjian, *Xinjian Xumuye*, Di san qi, pp. 59–60. {Chinese}
(Chen Xin, 2004, The key points in problem solving; practice of the process of discontinuation of livestock farming for grassland restoration and increase of grasslands, *Xinjang Xumuye*, The Third Period, pp. 59–60.)

Chi Yongming, 2004, Shengtai yimin shi xibu diqu shengtai huanjing jianshi de genben, *Jingji Luntan*, Di shiliu qi, pp. 14–15. {Chinese}
(Chi Yongming, 2004, Ecological migration as the base for the construction of ecological environment, *Economic Tribune*, The 16th Period, pp. 14–15.)

Dongribu, 2000, Shengtai yimin fupin de shijian yu qishi, *Zhongguo Pinkun Diqu*, 10 pp. 37–40. {Chinese}
(Dengribu, 2000, Practice and outcome of poverty alleviation by ecological migration, *China's Underdeveloped Regions*, 10, pp. 37–40.)

"Dongwu Kezue yu Dongwu Yixue", 2003, Neimenggu zizhiqu tuimuhuancao gongcheng jiri qidong, *Dongwu Kezue yu Dongwu Yixue*, Di san qi, p. 20. {Chinese}
("Animal Science and Veterinary Medicine", 2003, An enterprise of cultivation discontinuation for forest/grassland restoration executed in a short while in Inner Mongolia Autonomous Region, *Animal Science and Veterinary Medicine*, No. 3, p. 20.)

Du Ping, Zhang Zhuo & Hattori Kenji, 2004, Chugoku seibu daikaihatu no jisshi, *Chugoku 21 Tokushu Chugokuseibudaikaihatu*, (18), pp. 15–40. {Chinese}
(Du Ping, Zhang Zhuo & Hattori Kenji, 2004, Implemntation of "western development" policy, *China 21*, (18), pp. 15–40.)

Fan Hongzhong & Zhao Xiaodong, 2003, Xibu shengtai yimin wenti ji zhongdong diqu zai qizhong de zuoyong, *Nongcun Jingji*, Di qi qi, pp. 36–37. {Japanese}
(Fan Hongzhong & Zhao Xiaodong, 2003, Ecological migration issues in the west and their influence in the middle and east areas, *Chinese Rural Economy*, Vol. 7, pp. 36–37.)

Fang Bing, 2001, Jiada shengtai yimin lidu qieshi baohu xibu shengtai huanjing, *Xibu Dakaifa*, Di shisan juan, Di si qi, pp. 37–43. {Chinese}
(Fang Bing, 2001, Strengthen ecological relocation and protect western ecological environment, *West Development*, vol. 13, No. 4, pp. 37–43.)

Gao Chao, 2001, Shubianzhan damo tunken ying lüzhou – fang Xinjiang shengchan jianshe bingtuan silingyuan Zhang Qingli, *Xiaochengzhen Jianshe*, Di yi qi, pp. 10–14. {Chinese}

(Gao Chao, 2001, Border guards struggling with desert and staying in oasis – going on a visit to Zhang Qingli, commander of Xinjiang Production and Construction Corps, *Development of Small Cities & Towns*, No. 1, pp. 10–14.)

Ge Jianxiong, 1997a, *Zhongguo Yimin shi- Di yi juan,* Fujian renmin chubanshe. {Chinese} (Ge Jianxiong, 1997a, *History of Chinese Migration – Vol.* 1, Fujian People's Press.)

Ge Jianxiong, 1997b, *Zhongguo Yimin shi- Di er juan,* Fujian renmin chubanshe. {Chinese} (Ge Jianxiong, 1997b, *History of Chinese Migration – Vol.* 2, Fujian People's Press.)

Gegengaowa & Wuyunbatu, 2003, Neimenggu muqu shengtai yimin de gainian wenti yu duice, *Neimenggu Shehui Kexue*, Di ershisi juan, Di er qi, pp. 118–122. {Chinese} (Gegengaowa & Wuyunbatu, 2003, The Concept of ecological migration in the livestock farming district in Inner Mongolia, problems and provisions, *Inner Mongolia Social Sciences (Chinese Edition)*, vol. 24, No. 2, pp. 118–122.)

Guojiafagaiwei, Guojialiangshiju, Nongyebu deng Babuwei, 2003, "Tuimuhuancao he jinmu shesi chenhualiang gongying jianguan zanxing banfa" de tongzhi, *Neimenggu Xumu Kexue*, Di liu qi, pp. 79–80. {Chinese}
(Eighth Commission such as National Development and Reform Commission, State Grain Administration and Agricultural Department, 2003, A notification of 'The temporary measure for management of inedible grain supply subsidiary to cultivation discontinuation for grassland restoration, and, prohibition against pasturage and farming in feedlots', *A Journal of Animal Sciences and Production in Inner Mongolia*, No. 6, pp. 79–80.)

Hanhaisha, 2004, *Shachenbao Zhishi Wenda*, Zhongguo Shachen Wang, Zhongguo Qi-Xiangju, Gansusheng Qixiangju, <http://www.duststorm.com.cn/show.asp?ID=7251> (11yue 15hao). {Chinese}
(Hanhaisha, 2004, *Questions and Answers in Terms of the Knowledge of Sandstorm*, China Duststorm Net, The China Meteorological Administration, The Gansu Meteorological Administration, <http://www.duststorm.com.cn/show.asp?ID=7251> (November 15th).)

Hu Huazheng, 2004, *Shengtai Yimin de Ziyuan yu Feiziyuan Xing Yanjiu – Neimenggu Alashan meng Luanjingtan Diaocha*, Shuoshi xuewei lunwen, Wu yue yi hao, Zhongyang Minzu Daxue. {Chinese}
(Hu Huazheng, 2004, *Studies of Voluntariness and Involuntariness in Ecological Migration – Investigation into Alasha Luanjingtan League in Inner Mongolia*, Master's Thesis, 1st May, Submitted to Central University for Nationalities.)

Huang Weixiong, 1987, Renkou pingheng yu shengtai xitong – Jianlun xiang xibei yimin wenti, *Foshan Kexue Jishu Xueyuan Xuebao (Shehui Kexue Ban)*, Di yi qi, pp. 67–75. {Chinese}
(Huang Weixiong, 1987, A balance of population and an ecological system – in addition, an argument about problems of migration in the northwest, *Journal of Foshan University (Social Science Edition)*, Vol. 1, pp. 67–75.)

Li Ning & Gong Shijun, 2003, Lun Ningxia diqu shengtai yimin, *Haerbin Gongye Daxue Xuebao (Shehui Kexue Ban)*, Di wu qi, Di yi hao, pp. 19–24. {Chinese}
(Li Ning & Gong Shijun, 2003, On the ecological migration in Ningxia, *Journal of Harbin Institute of Technology (Social Sciences Edition)*, Vol. 5, No. 1, pp. 19–24.)

Li Xiaochun, Chen Zhi, Ye Liguo, Dong Hua, Liu Min, Zhang Jun and Nie Fuling, 2004, Dui shengtai yimin de lixing sikao – Yi Hunshanke-shadi wei li, *Neimenggu Daxue Xuebao (Renwenshehuikexueban)*, Di san shi liu juan, Di wu qi, pp. 34–38. {Chinese}
(Li Xiaochun, Chen Zhi, Ye Liguo, Dong Hua, Liu Min, Zhang Jun and Nie Fuling, 2004, Rational reflections on "ecological migration", *Journal of Inner Mongolia University (Humanities and Social Sciences)*, Vol. 36, No. 5, pp. 34–38.)

Liu Bingfeng, 2004, Mao Zedong yu gongheguo "Tunken shubian" shiye, *Zhongguo Jingniu Mao Zedong Wang*, <http://www.mzdthought.com/wz/110zn/mzdyghg.htm> (12yue 31hao). {Chinese}
(Liu Bingfeng, 2004, Mao Zedong and republic's project to 'Protect the border together with cultivation', *China Jingniu Mao Zedong's Thought Net*, <http://www.mzdthought.com/wz/110zn/mzdyghg.htm> [December 31st].)

Liu Jun, 2000, Alashan – Tuimuhuancao suo shachen, *Renmin Ribao Haiwaiban*, 7yue 13hao, Di shiyi mian. {Chinese}
(Liu Jun, 2000, Alasha – Cultivation discontinuation for grassland restoration in relation to cloud of sand, *People's Daily International Edition*, July 13th, the 11th Page.)

Liu Xuemin, 2002, Xibei dequan shengtai yimin de xiaoguo yu wenti tantao, *Zhongguo Nongcun Jingji*, Vol. 4, pp. 47–52. {Chinese}
(Liu Xuemin, 2002, On the effect and problem of ecological migration in the Northwest, *Chinese Rural Economy*, Vol. 4, pp. 47–52.)

Liu Xuemin & Chen Jing, 2002, Shengtaiyimin chengzhenhua yu chanye fazhan- Dui xibeidiqu chengzhenhua de diaocha yu sikao, *Zhongguo Tese Shehuizhuyi Yanjiu*, Di er qi, pp. 61–63. {Chinese}
(Liu Xuemin & Chen Jing, 2002, Ecological migration, urbanisation and industrial development – research and consideration in terms of urbanisation in the northwest, *Studies on the Socialism with Chinese Characteristics*, No. 2, pp. 61–63.)

Liu Zenglin, 2002, Wei le minzu diqu de caofengshuimei – Quanguo zhengxie minzongwei tichu "Tuimuhuancao" jianyi jishi, *Zhongguo Minzu*, Di wu qi, pp. 38–39. {Chinese}
(Liu Zenglin, 2002, For the rich grasslands and sweet water in ethnic minority regions – a report about proposal of 'Cultivation discontinuation' submitted by Chinese People's Political Consultative Conference, *China Ethnicity*, No. 5, pp. 38–39.)

Luo Kanglong, 1993, Mingqing liangdai Guizhou hanzu yimin tezheng de duibi yanjiu, *Guizhou Shehuikexue*, Di san qi, pp. 104–108. {Chinese}
(Luo Kanglong, 1993, Comparative analysis on characteristics of Han migration into Guizhou in Ming and Qing Dynasty, *Social Sciences in Guizhou*, No. 3, pp. 104–108.)

Meng Linlin, 2004, *Shengtaiyimin dui Mumin Shengchanshenghuo Fangshi de Yingxiang Yanjiu – Yi Aolike-Gacha wei li*, Shuoshi xuewei lunwen, 4yue1hao, Zhongyang Minzu Daxue. {Chinese}
(Meng Linlin, 2004, *On the Influence of Eco-migration to Herdsmen's Producing and Living Patterns: A Case Study of Aolike Gachaa*. Master's Thesis, 1st April, Submitted to Central University for Nationalities.)

Meng Linlin & Bao Zhiming, 2004, Shengtaiyimin yanjiu zongshu, *Zhongyang Minzu Daxue Xuebao (Zhexue Shehui Kexue Ban)*, Di san shi yi juan, Di liu qi, pp. 48–52. {Chinese}

(Meng Linlin & Bao Zhiming, 2004, Summarization of the study of migration for the reason of zoological environment. *Journal of The Central University for Nationalities (Philosophy and Social Sciences Edition)*, Vol. 31, No. 6, pp. 48–52.)

Onishi Yasuo, 2004, Chugoku seibu daikaihatsu no hyoka to tembo, *Chugoku 21*, (18), pp. 41–56. {Japanese}

(Onishi Yasuo, 2004, The evaluation and prospect of China's western development strategy, *China 21*, (18), pp. 41–56.)

Pi Haifeng, 2004, Xiaokang shehui yi shengtaiyimin, *Nongcun Jingji*, Di liu qi, pp. 58–60. {Chinese}

(Pi Haifeng, 2004, A stable society and ecological migration, *Rural Economy*, No. 6, pp. 58–60.)

Quanguo renmin daibiao dahui changwu weiyuanhui, 1989, Zhonghua renmin gongheguo huanjing baohu fa. {Chinese}

(The Standing Committee of the National People's Congress, 1989, Environmental protection law of the People's Republic of China.)

Quanguo renmin Daibiao Dahui Changwu Weiyuanhui, 1998, Zhonghua renmin gongheguo senlin fa. (Yuanan 1984 nian) {Chinese}

(The standing committee of the National People's Congress, 1998, Forest law of the People's Republic of China (First draft: 1984).)

Quanguo renmin daibiao dahui changwu weiyuanhui, 2001, Zhonghua renmin gongheguo fangsha zhisha fa. {Chinese}

(The Standing Committee of the National People's Congress, 2001, "Law of the People's Republic of China on prevention of sand invasion")

Quanguo renmin daibiao dahui changwu weiyuanhui, 2002, Zhonghua renmin gongheguo caoyuan fa. (Yuanan 1985 nian) {Chinese}

(The Standing Committee of the National People's Congress, 2002, Grassland law of the People's Republic of China(First draft: 1985).)

Ren Yuewu, Yuan Guobao & Li Fenghu, 1993, Shilun Sanxia-kuqu shengtaiyimin, *Nongye Xiandaihua Yanjiu*, Di shi si juan, Di yi qi, pp. 27–29. {Chinese}

(Ren Yuewu, Yuan Guobao & Li Fenghu, 1993, On the ecological migration in Sanxia reservoir region; *Research of Agricultural Modernization*, Vol. 14, No. 1, pp. 27–29.)

Sang Minlan, 2004, Lun Ningxia de "shengcunyimin" xiang "shengtaiyimin" de zhanlüe zhuanbian, *Shengtai Jingji*, S yi qi, pp. 23–25. {Chinese}

Sang Minlan, 2004, On the strategic change from "subsistence migration" to "ecological migration" in Ningxia Autonomous Region, *Ecological Economy*, No. S1, pp. 23–25.

Wang Daming & Yan Hongbo, 2001, Tuimuhuancao – gaishan caodi shengtai huanjing, *Qinghai Caoye*, Di shi juan, Di san qi, pp. 37–39. {Chinese}

(Wang Daming & Yan Hongbo, 2001, Improvement of ecological environment in rangeland by de-stocking, *Qinghai Prataculture*, Vol. 10, No. 3, pp. 37–39.)

Wang Ke, 2001, "Shosu minzoku" kara "kokumin" he no dotei – gendai chugoku ni okeru kokumintogo to iu shiten kara, *Ajia Kenkyu*, 47(4), pp. 39–62. {Japanese}

(Wang Ke, 2001, From "minority ethnic groups" to "national citizens" – the unification process in modern China, *Asian Studies*, No. 47, Vol. 4, pp. 39–62.)

Wang Peixian, 2000, Shengtai-yimin – Xiaochengzhen jianshe yu xibu fazhan, *Guotu Jingji*, Di liu qi, pp. 25–26. {Chinese}
(Wang Peixian, 2000, Construction of small cities and development in the west, *Territory Economy*, Vol. 6, pp. 25–26.)
Wang Xiangyang, 2003, Zhongguo xibu muqu tuimuhuan de zhengce zhichi, *Nongye Jingji Wenti*, Di qi qi, pp. 45–50. {Chinese}
(Wang Xiangyang, 2003, The policy support of grazing forbidden in western pastoral, *Issues in Agricultural Economy*, No. 7, pp. 45–50.)
Wu Songdi, 1997a, *Zhongguo Yimin shi- Di san juan*, Fujian Renmin Chubanshe. {Chinese}
(Wu Songdi, 1997a, *History of Chinese Migration – Vol. 3*, Fujian People's Press.)
Wu Songdi, 1997b, *Zhongguo Yimin Shi- Di si juan*, Fujian Renmin Chubanshe. {Chinese}
(Wu Songdi, 1997b, History of Chinese Migration – Vol. 4, Fujian People's Press.)
Wulantuya, 2000, A-meng shengtaiyimin bandechu wendezhu tuopin kuai, *Neimenggu Ribao (Han)*, 10yue 31hao, Di si mian. {Chinese}
(Wulantuya, 2000, Alashan-league Ecological Migration Migrate, Settle Down and Get Away from Poverty in a Short Time, *Inner Mongolia Daily (Chinese Edition)*, October 31st: The Fourth Page.)
Wuligeng, 2003, Shilun shengtaiyimin gongzuozhong de minzuwenti, *Neimenggu Shehui Kexue*, Di er shi si juan, Di si qi, pp. 12–14. {Chinese}
(Wuligeng, 2003, On the ethnic issues in practice of ecological migration, *Inner Mongolia Social Sciences (Chinese Edition)*, Vol. 24, No. 4, pp. 12–14.)
Xie Guoxian, 1996, Mingdai Yunnan de Hanzu yimin, *Yunnan Minzu Xueyuan Xuebao (Zhexue Shehui Kexue ban)*, Di er qi, pp. 24–30. {Chinese}
(Xie Guoxian, 1996, Han migration into the Yunnan Area in Ming Dynasty, *Journal of Yunnan Institute of the Nationalities (Social Sciences Edition)*, No. 2, pp. 24–30.)
Xu Feng, 2003, Neimenggu Tuimuhuancao buzhu biaozhuo queding, *Caodi Kexue*, Di ershi juan, Wu qi, p. 56. {Chinese}
(Xu Feng, 2003, Determination of the standard for aid to cultivation discontinuation for grassland restoration in Inner Mongolia, *Pratacultural Science*, Vol. 20, No. 5, p. 56.)
Xu Honggang, 2001, "Shengtaiyimin" zhengce dui huanjie caoyuan shengtai yali de youxiaoxing fenxi, *Guotu yu Ziranziyuan Yanjiu*, 4, pp. 24–27. {Chinese}
(Xu Honggang, 2001, Analysis on eco-migrate policy efficiency in grassland ecopress relaxation, *Territory & Natural Resources Study*, No. 4, pp. 24–27.)
Xu Suhuan, 2003, Xinan shanqu "Tuigenghuanlin" yu "Shengtaiyimin" waiqian yanjiu, *Xibudakaifa yanjiu*, Di san qi, pp. 116–119. {Chinese}
(Xu Suhuan, 2003, Investigation into migration subsidiary to 'reforestation of cultivated land' and 'ecological migration' in the Southwest Mountains, *Investigation into West Development*, No. 3, pp. 116–119.)
Yin Yue and Chai Hailiang, 2002, Neimenggu jihua touzi shang yi yuan liunian shengtaiyimin liubai wushi wanren, *Renmin Ribao Haiwaiban*, 12yue2hao. {Chinese}
(Yin Yue and Chai Hailiang, 2002, An investment of hundred million yuan in order to generate 650,000 Ecological migrations in a 6-year period in Inner Mongolia, *People's Daily International Edition*, December 2nd.)

Zheng Chao, 2000, Shennongjia shouchuang shengtaiyimin, *Jilin Huanjing Bao*, 7yue2hao, Di er mian. {Chinese}
(Zheng Chao, 2000, Shennongjia created ecological migration, *Jinlin Environment Paper*, July 2nd: The 2nd Page.)

"Zhongguo Muye Tongxun", 2003, Tuimuhaicao gongchen quanmian qidong, *"Zhongguo Muye Tongxun"*, 2 yue B ban, p. 10. {Chinese}
(China Animal Husbandry Bulletin, 2003, Entire invocation of the process of cultivation discontinuation for grassland restoration, *China Animal Husbandry Bulletin*, February B Edition, p. 10.)

Zhonghua Renmin Gongheguo Guowuyuan, 2002, Tuigenhuanlin tiaoli. {Chinese}
(The State Council of the People's Republic of China, 2002, The regulations on restoring farmland to forest.)

*I Questioning ecological aspects?
Can "ecological migration" result
in environmental conservation?*

1 The beginnings of "ecological migration" in the Heihe River valley from case studies in Ejene Banner, Alasha League, Inner Mongolia Autonomous Region

YUKI KONAGAYA

Introduction

We first visited the Heihe River valley in August 2000. In order to undertake an anthropological study of changes in nomadic society in the Mongolian tablelands – funded by a Grant-in-Aid for scientific research – we traveled through the Qinghai region, then north from Jiuquan to Ejene banner in the Alasha (Alashan) League of the Inner Mongolian Autonomous Region. We were accompanied on our survey by Annette Erler, a German ethnomusicologist based at the Royal Danish Museum in Copenhagen, engaged in research on Mongolian ethnic music. For her, Ejene is a special place. The museum she works for houses an extensive collection of Mongolian ethnic musical instruments and recordings from the early 20th century, some of which were originally obtained in Ejene Banner. These items were first collected and analyzed by Henning Haslund-Christensen and his colleagues others who took part in an expedition to the area led by Sven Hedin, a renowned explorer of inland Asia, between 1927 and 1930. For Annette Erler, the main purpose of this trip was to collect folk songs that have survived to this day by oral tradition.

We joined Annette on one of her visits to an elderly woman who had been introduced by the local education committee as someone who was well versed in the art of folk singing. Before enquiring whether she would sing for us, we listened to her story. (Photo 1-1)

> Long ago, there were many poplar trees. And in the poplar trees there were cuckoos that sang. When the cuckoos sang a lot, it would rain. The reeds around the rivers grew so high that we could not see the camels behind them. Today, the land is still there, but there is no grassland anymore. There is only thin grass, like moss. Long ago there were many plants, but now there are only poplars, tamarisk and goosefoot.

Photo 1-1. Elderly lady in Ejene, recounting her story
(Photo by Yuki Konagaya, August 2000)

These days the sun beats down on the earth, rain rarely falls, and the land is plagued by sand. Yet, the elderly ladies insist that there was once forest inhabited by cuckoos. Sven Hedin too. In an account of his travels, "Across the Gobi Desert," he describes this place as "an earthly paradise" (Hedin, 1931: 182) So, what really happened to this land? What factors contributed to transforming this oasis to a semi-desert in just half a century?

Elucidating what changes have occurred in the water resources of this area during the 20th century will undoubtedly provide important clues for understanding the path followed by the inhabitants of this region to date and also for plotting an appropriate future course. It was based on this premise that a new research project was launched in 2001. The project was one of several undertaken by the Research Institute for Humanity and Nature and was titled, "Historical evolution of adaptability in an oasis region to water resource changes" (also referred to as the Oasis Project). Full-scale surveys were initiated in 2002.

The research paper presented here is based on field notes taken during surveys conducted with Tomohiro Akiyama in June 2002 as part of the Oasis Project. Mr. Akiyama is a young hydrology researcher specializing in the groundwater dynamics of arid environments. In order to measure groundwater levels, we decided to conduct surveys at various households and inspect the wells that the local inhabitants used for potable water. Measuring the water level of a well may initially appear to be a simple task, but in reality it is often considerably difficult. This is because many of the household wells in the area have been converted into pumping wells and open wells from which water is drawn manually are no longer frequently found. Without asking the local inhabitants it would be impossible to locate the more traditional, open wells, or which households were still using such wells. Our surveys began by adopting the local people as our teachers and guides, which enabled us to perform simple groundwater measurements by dropping lines down into the wells to measure their depth. As they spoke to us, the anxieties of the people here became apparent.

At this time, the policy of "ecological migration" was finally about to be implemented. Whenever people gathered, they expressed vague fears about their future, about who would be moved to where, about forced removals, and about whether they would be able to sustain themselves after moving. In this paper, I would like to document the distress of these people at that time. In a sense, this paper describes the situation of the at the lower reaches of the Heihe River on the "eve" of "ecological migration."

When the policy generally referred to as "ecological migration" was fully implemented after 2002, the fears of the people became a reality. Of course, in some cases people conquered their fears and went on to achieve success. In fact, the measures associated with the promulgation of the "ecological migration" initiative in the Inner Mongolia Autonomous Region have generally come to be regarded as a model and have set the precedent for further expansion of the policy throughout the country. In this sense, the conditions associated with the "eve" we witnessed previously in Ejene Banner are likely to be repeated in other areas of China. It is thus hoped that by documenting this "eve" we may gain valuable insights that will help to identify problems at an early stage, in order that the policy will be implemented appropriately.

The flow of the Heihe River over 2000 years

The Heihe River has its origins in the glacial region of the Qilian Mountains, which extend east to west into northwestern China. The river carries water to oasis cities such as Zhangye and Jiuquan before flowing further north where it forms lakes in the desert near the border with Mongolia. The river is 400 km in length and has a total catchment area of 140,000 km^2, making it China's second largest inland river after the Tarim (Fig. 1-1).

The lake into which the Heihe empties was historically referred to as the *Juyanze* or *Juyanhai* in the literature. Land clearance by ex-legionnaires dated back to the Han period, and many writings on wood – known as *Juyanhanjian* – have been excavated in the area. The existence of extensive historical records like these in an area so very remote from the so-called "cultural center" of China reveal that the region was cleared for agriculture by peasants long ago and that it also served as an invasion route for nomadic herders. Based on these historical references it is thus quite apparent that the area was of great strategic importance to the rulers of the time.

This area of ancient development corresponds to the lower reaches of the Heihe River, which was also called the *Ruoshui* ("weak waters"); a name that may have referred to the relative ease with which inhabitants controlled and utilized the water of the river. Irrigation infrastructure was developed and the land was cleared for cultivation. According to Mitsuyuki Inoue, who is responsible for deciphering ancient literature in the Oasis Project, the 3rd century-BCE chronicle "*Houhan shu*" states that there were 1,560 homes and a population of 4,733 in the area. Although the credibility of these figures may be questionable, it is apparent that there was considerable development in the area.

In addition to engaging in agricultural activities, the *tuntianbing* (soldiers of military-farming colonies) were also entrusted with preventing invasions from the north by nomadic herders on horseback. In order to detect the routes of horseback nomads who would advance under the cover of darkness, the *tuntianbing* leveled the sand to make footprints easily discernible (Momiyama, 1999).

The nomadic people enjoyed overwhelming military superiority by virtue of their large numbers of horses and other livestock, and were unfamiliar with any form of trade negotiation that did not involve the use of force. From the perspective of the defenders this approach would have constituted being plundered, but to the small groups of these nomads that were active in a region where there were few places that were probably worth attacking, the lower reaches of the Heihe must have presented a worthwhile target.

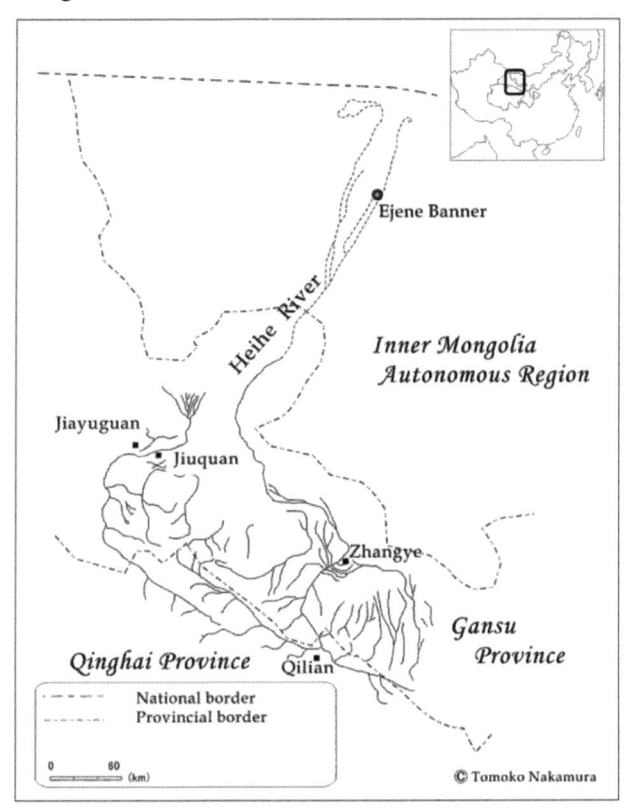

Fig. 1-1. Map of the Heihe River Basin

In the 11th century this area came under the rule of the Shishia, and later, by the Yuan dynasty. The ruins of the castle city at this time, *Khara khoto*, meaning "black town" in Mongolian, is now a tourist attraction. According to Noriyuki Shiraishi, who is responsible for archeological surveys in the Oasis Project, the size of the divisions of the ruin suggest that the castle was originally constructed in the Han period and that it was from that time. Given the considerable size of the surrounding area measuring of construction, it has also been proposed that the area was continuously inhabited through the centuries.

However, by the latter half of the 14th century the quantity of water in the old lake (Juyanze) had fallen to just one third of its previous capacity, and by the late 15th century the castle city was abandoned. It is thought that one of the reasons for the decline of the town was the scarcity of water and elucidating the underlying cause of this drying up is one of the principal tasks of the Oasis Project. Whatever the reason, the main course of the river has shifted westward since that time and today the Heihe River runs approximately 30 km west of where it flowed when it was known as *Ruoshui* ("weak water"). Despite the fact that the area is extremely arid with less than 50 mm of precipitation annually, the presence of the river at that time supported a diversity of flora and created a green corridor along the river consisting of trees such as poplars (a member of the deciduous Salicaceae order; in Chinese *huyang*; in Mongolian *toorai*; genus *Populus euphratica*) and ash trees (deciduous arbor of the order Elaeagnaceae; in Chinese *shazao*; in Mongolian *jigde*; genus *Elaeagnus amgustifolia*).

In the 18th century the area underwent a major change. In 1704, the Torgud group of Oirad Mongolian origin obtained permission from the Qing dynasty to settle in the area where they became nomadic herders. The group is said to have migrated to the western shores of the Caspian Sea in 1630 after which, due to their unwillingness to convert to Greek Orthodoxy, they crossed the Volga and returned once more to the east. According to the traditional beliefs as recorded in the "Ejinaqizhi" anthology and other literature, the group cleared the forests lining the river in order to create grasslands, suggesting that the group of Mongolian herders had settled in the green corridor.

The course of the Heihe River has shifted substantially from east to west, and its lower reaches were transformed from being used as agricultural land into use for grazing. However the strategic importance has persisted to the 20th century. In the 1930s, the Japanese established a special military agency here, and in 1958 the People's Liberation Army was stationed in the area after forcing the inhabitants to emigrate from the area. Given the strategic importance and proximity of the international border, a large military base was established in the 1960s. While the military value of this area has remained unchanged, the second half of the 20th century has brought with it considerable change in the distribution of water resources.

A major construction project was initiated in 1957; the construction of the Yingluoxia Dam in the upper reaches of the Heihe River and the Zhengyixia floodgates in the middle reaches of the river. Together these developments resulted in the promotion of agricultural development on a wide scale and ushered in the era known as the "Great Leap Forward." In the upper reaches of the Heihe, extensive areas of forest were felled and large numbers of migrants were settled in the area from the Tibetan Plateau and other areas. The middle reaches of the river too were settled with migrants and agricultural activities and the land were expanded. At successive sites located in the lower river reaches, urban centers were constructed, nomadic herders were settled, Han migrants came to live and to cultivate.

At each of the upper, middle and lower reaches of the Heihe, increasing expanses of land were reclaimed and the population increased dramatically, becoming particularly concentrated along the middle reaches of the river (Fig. 1-2). For example, the population of Zhangye was 220,000 in 1949, it exceeded 300,000 by 1957 and was 430,000 by 1990. Along the middle reaches, where the carrying capacity is highest, reservoirs were constructed in the western areas in 1970. However, subsequent to the construction of these reservoirs the flow of water (originating at a different source to the Heihe) into Lake Beidahe in Ejene Banner get fewer. Over the second half of the 20th century, development along the middle reaches had the effect of increasing water demand to such an extent that the lower reaches dried up. One of two lakes located in the lower reaches of the Heihe River, Lake Sogonoor, dried up completely in 1992 and has subsequently almost vanished. (Fig. 1-1)

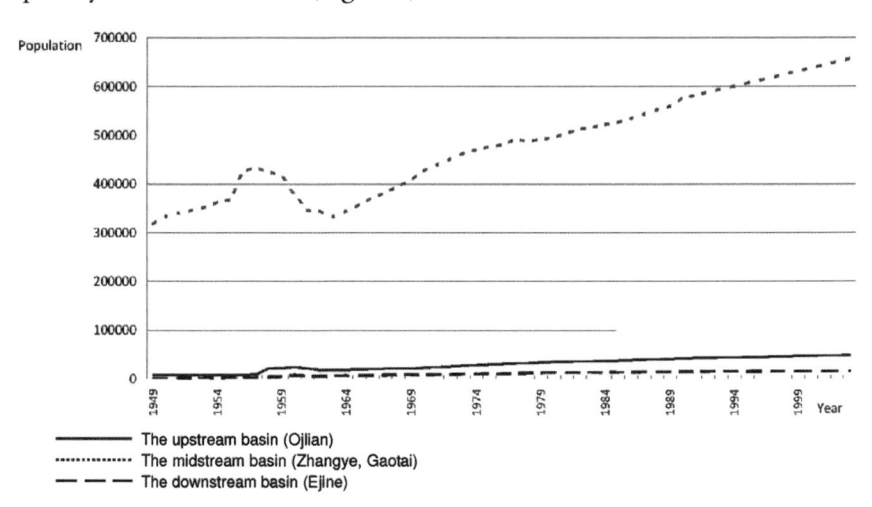

Fig. 1-2. Population growth along the Heihe River
© Tomoko Nakamura

Table 1-1. Change in area of Lake Sogonoor

Year	1950s	1960s	1970s	1980s
Area (km^2)	302	35.5	30	23.6

Source: Created by Tomoko Nakamura based on Wang and Cheng 2000: 787.
G. Wang, G. Chang 2000, 'The characteristics of water resources and the changes of the hydrological process and environment in the arid zone of northwest China', Environmental Geology 39 (7) May.

In 1979, when the lower reaches of the Heihe were incorporated into the Inner Mongolia Autonomous Region, it became necessary to distribute water resources beyond the borders of Gansu Province in which the middle reaches of the Heihe are located, water resource management became complicated. Finally, in 1992, the National Irrigation and Drainage Department set up a special committee to draft legislation related to the distribution of water. The committee decided that the quantity and timing of water released into the lower river reaches would be determined through consultation when necessary. At present, the district of Zhangye has become a model for the promotion of water-saving irrigation, and, as a matter of policy, considerable attention is focused devoted to the management of water resources in the lower reaches.

As evidenced by the slogan, "Building a water-saving society," in China today, sustainable utilization of natural resources and environmental conservation are extolled as being virtues associated with development. However, just as efforts directed at environmental conservation were gaining momentum, the government embarked on a policy of "ecological migration" in 2001.

The concept of conserving the forests in the upper reaches of the Heihe River is referred to as "water source cultivation", which is important for water retention. For this reason, the idea of persuading herders to migrate from the upper reaches – to alleviate the pressure associated with grazing by goats and other animals responsible for damaging forests in water catchment areas – became a matter of policy. Migrant villages were then constructed around the middle reaches of the Heihe to accommodate these migrants, which increasing the considerable investment even further. This form of development is referred to as "water-saving irrigation." At the lower reaches, it was considered necessary for people living along the river to move further away from its banks to facilitate "ecological recovery" of the poplar forests that were decreasing due to lower groundwater levels. It has been through the promulgation and execution of policies such as these that people have been relocated for the sake of conserving the ecological environments between the lower and upper reaches of the Heihe River.

The "eve" of "ecological migration" in the lower Heihe

More than any other reach along the length of the Heihe, the lower reaches were the first to show signs of the great contradiction that would become associated with the policy of "ecological migration". It's usually told that the storms of swirling sand that blanketed Beijing were the result of the withering riparian poplar forests and the vanishing of the lake. But it is clear that these events were due to the river drying up, and not due to the activities of the area's inhabitants. Nonetheless, to protect the poplar forests, a quota of 1,500 inhabitants was earmarked for migration (from a total of 514 households) from the area within a three-year period from 2001. According to local administration officials, migration to three types of destination are planned: to migrant villages constructed on the outskirts of the banner "capital", migrant villages in areas around livestock feed bases built on land cleared for agriculture, and to remote areas for camel herders.

While we were conducting our preliminary surveys in the August and September of 2001, the people of the area began referring to the "ecological migration" policy as the "second era of forced migration." Many people in the area had already experienced the forced migrations in 1957, when they were relocated from Bayanbogd County to the south to make way for the construction of a large military base. Their recollections of that time reverberated throughout the community. By June 2002, while we were conducting additional surveys, the selection of inhabitants who would be relocated from the area under the "ecological migration" policy was supposedly being finalized. However, the majority of inhabitants did not want to move. Next, I will present several case studies of inhabitants who employed a variety of inventive means to use the natural resources in order to sustain their livelihoods, and how, irrespective of these developments, they were forced to abandon these methods of resource utilization (Fig. 1-3).

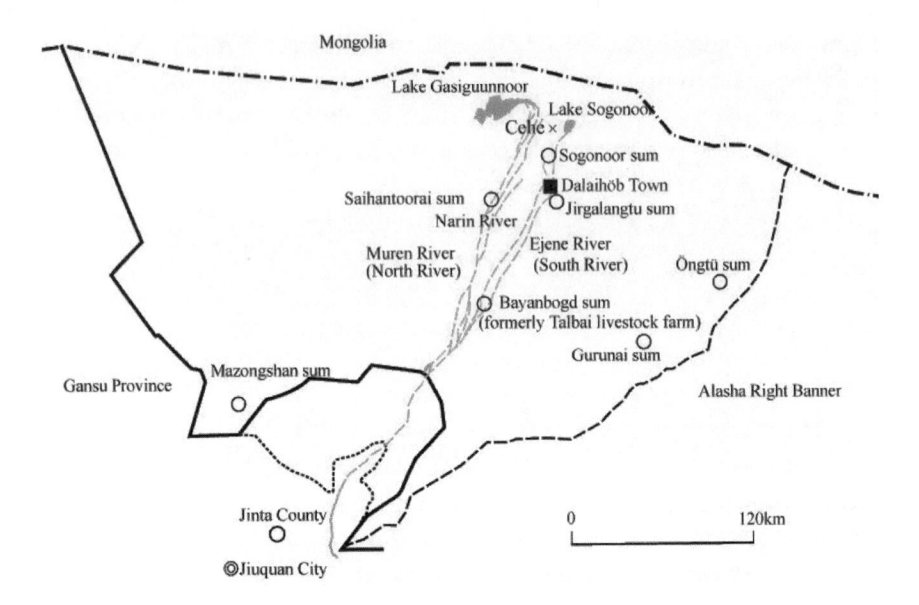

Fig. 1-3. Map of Ejene Banner
© Kanako Kodama

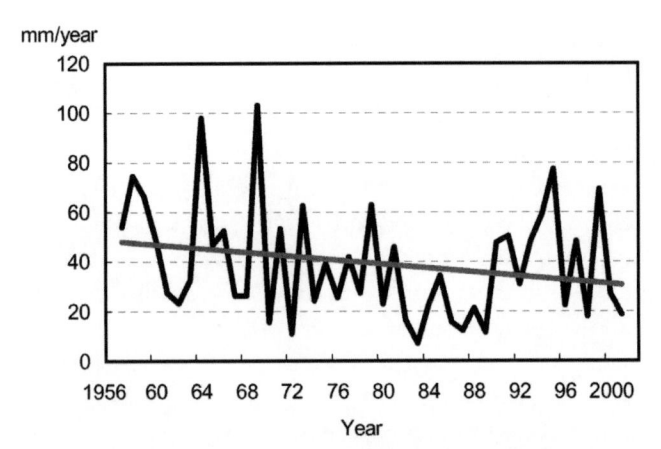

Fig 1-4. Annual precipitation in Ejene Banner
Tomohiro Akiyama

Note: Based on materials from the Ejene Banner Weather Bureau

Features of the natural environment

The annual precipitation in Ejene banner is as low as 40 mm (Fig. 1-4). Although rainfall has decreased over the past 50 years, it has done so gradually; even in the past the area was extremely arid. The belt of green that flourished here, despite this aridity, was due to the substantial river runoff from the Heihe River that has its origins in the Qilian Mountains.

At the point where it enters from Gansu province into Ejene banner, the Heihe River splits into the Ejene River in the east that empties into Lake Sogonoor (local name: Sob nanoor) and the Muren River in the west that empties into Lake Gasiguunnoor. In Mongolian, *muren* means "wide river," and along its course a number of branches split off and rejoin the main river course before the river reaches its end in the lake in the middle of the desert. The course of another river, the Narin River (meaning "narrow river") flowed in parallel with the Ejene River. Like the Ejene, it also flowed into a lake Sogonoor and formed a delta.

From the west to the east, the Muren, Narin and Ejene Rivers all generally flowed in the same direction, running from SSW to NNE. People here use the rivers to confirm their bearings, considering the left bank as their "north" and the right bank as their "south." That is, with WNW behind them (facing ESE) they take the direction in front of them as "south" and that behind them as "north."

This is a slightly different folk perception of the cardinal directions compared to those traditionally in Mongolia where southeast is regarded as the "front" and "south," and northwest is regarded as "back" and "north". It is closer to the system employed by the Mongolian Torgud people, who consider east as being "south" and west as "north." It might be that this system reflects the perception of the physical geography of the region by the Torgud given that they returned from the far west of the continent. Also, from a more practical standpoint, the fact that winter winds blow exclusively from the west or WNW makes it desirable to erect tents with their backs to the winds.

Utilization of the natural environment

Formerly, poplar forests grew along the rivers. Along the Muren, which was characterized as having consistently high runoff due to relatively high rainfall, groves of ash were also found. In the central area of the banner, there were several lakes and marshes where reeds (perennial herbs of the order Gramineae; in Chinese *luwei*: in Mongolian *hulusun*; genus *Phragmites australis*) also flourished. The indigenous name of the central area of the banner is Dalaihöb, meaning "depths of the sea." A woman born in 1930 near Lake Muhur, located a little upstream of Dalaihöb, describes the landscape here in the 1950s as follows:

> There were many lakes around this town. Wild geese and ducks and other migratory birds came here. Since swans come to another east lake, not many are seen around here anymore. Due to the change in water levels we have sometimes seen fish dying. At these times, people scooped up the fish and returned them to deeper water. The river water never stopped flowing. In winter the water froze, and in spring it melted and the water increased. The water level then gradually fell, but in June it would rain again, which meant that water levels would increase again in summer. When I went to the shores of the lake there were many shellfish and spiral shells and I often played with them. The reeds growing at the shores of the lake were so high that we could not see the camels behind them.

The natural environment of this area was thus once characterized as having rivers that flowed through the desert. It is for this reason that people use the word *gol* to describe, not only the rivers, but also the entire natural environment along the river banks. They also referred to the desert *gobi*, which is topographical term, and a region in the desert as *köl* which is also topographical term. There, in the area of desert around the river that once flowed into the Juyanze River before it vanished, reeds can still be found and groundwater levels are high. At another place too, with the indigenous name *Gurunai*, it is said that streams of water can periodically be seen.

It is known that the people of the area once adopted a general pattern of seasonal movement – in summer people moved to *gobi* to enjoy the breeze and stay cool and in winter they went to *gol* to avoid the wind and be warm. This "summer in *gobi*, winter in *riverine*" pattern was particularly prevalent in the People's Commune Era (1958–1983). This movement was also restricted as livestock grazing was barred in the area along the river until the end of the harvest season.

Before the People's Commune Era, wealthier people who owned large animals such as camels and horses wintered in places more remote as *köl* and *Gurunai*, as they could travel long distances easily. Thus, in contrast to the "summer in *gobi*, winter in *gol*" pattern described previously, some people adopted the custom of returning to the river district in summer.

During the People's Commune Era this seasonal migration pattern continued to be practiced by people employed in grazing camel herds.

There was thus once a seasonal pattern of moving to and away from the river, and the powerful meaning of the river course as a reference for spatial perception was closely integrated with the behavior of the people.

With the dissolution of the people's commune system in the area in 1983, livestock and grazing land were distributed among the inhabitants and there was a dramatic increase in the rate of settlement. As the inhabitants sold the camels that were distributed among households due to the difficulties associated with managing them, they lost their means of transport. In addition, the deterioration of natural grassland prompted the inhabitants of *gol* (riverine) to start growing livestock feed crops instead, which meant that they were no longer able to leave *gol* at any time of the year. Since 2002, people only travel distances ranging between a few hundred meters to approximately 10 km, with an average of between one to three kilometers. There has thus been a marked shift to a sedentary lifestyle where people live in fixed dwellings and they move within a small area to graze their animals.

Associated with the loss of a nomadic lifestyle has been an increased difficulty to adapt to changes in weather, particularly among livestock. For this reason, people with access to transport employ a method known as *otor*, by which livestock is grazed in areas far from the settlement and only the persons responsible for attending the animals go on these trips. Especially in times of drought, animals are taken to the rich natural environment of Mt. Honggor near the Mongolian border.

Recommendations to migrate

Upper reaches of the Muren: Compared to the "south river", water has historically been more abundant in the "north river" (Muren) which had groves of fruiting ash trees along its banks. However, in 1952, the seat of the banner government was shifted from the town of Saihantoorai at the lower reaches of the "north river" to Dalaihöb along the "south river." Since that time, water from the upstream has been arranged much to the "south river" than to the "north river". The increased runoff as a result of the transfer has resulted in the establishment of the ash trees that were planted along the banks of the "south river" in the latter half of the 20th century.

The house of family A at the upper reaches of the Muren is brand new. When they were allocated their land, they had a winter campsite along the river and a summer campsite at a higher location. The distance between the two campsites was less than two kilometers. The family stayed at the winter campsite between November and April. However, the erection of fencing to protect the poplar trees along the river meant that the family could no longer use their winter campsite and forcing them to cease such trips and to settle at their summer campsite. Half the construction costs for their new house came from a government subsidy. While the family has thus successfully managed to secure themselves a house, they are still uncertain about how they will manage their livestock. For now, they are thinking about planting grass to produce fodder.

Since grazing was banned in certain areas of the redistributed land, the family received financial compensation for not abandoning the area and settling elsewhere. They were not forced to migrate to a totally different area in this particular case as water is relatively abundant in the upper river reaches and, consequently, "ecological migration" is not enforced here. However, the seasonal practice of dividing their time between *gol* and *gobi*, as a traditional form of environmental utilization is no longer possible.

Lower reaches of the Muren: The administrative district of Bayantal (meaning "rich plains") was established in the lowest reaches of the river where it empties into Lake Gasiguunnoor. During the People's Commune Era, this area was used as grazing lands for horses.

Until the 1960s, the area consisted of green plains without any tamarisks and was dotted by houses separated by an average distance of approximately five kilometers. In winter, the people stayed in riverfront. In summer, as the grasses began to bud, the inhabitants returned to what they referred to as the "white place" (meaning open spaces). In the 1970s, the flow of water in the Muren River decreased, which prompted people to begin digging wells.

Ms. B lives within the administrative district, at the corner of a settlement that is equipped with an electric water pump that operates around the clock. The number of empty houses here is striking. There are, reportedly, only three households left in the settlement. The family next door moved to Mazongshan, 150 km away. Mazongshan is one of the destinations to which ecological migrants can move. The destinations offered to Ms. B are Saihantoorai in the central area of the banner and upstream and along the banks of the Muren River. The other is Hongxing, a livestock farming settlement on the outskirts of Saihantoorai. Given that she does not wish to choose either of these options, Ms. B has remained in her house, hesitating. She says that even if water is released into Lake Sogonoor soon, no water will be released into Lake Gasiguunnoor.

This is one example in which a political decision was taken to cease the flow of water to a particular area. In other words, in order to protect Dalaihöb, the most important town in Ejene Banner, in the future, water supply policy will favor the Ejene River and Lake Sogonoor in the east over the Muren River and Lake Gasiguunnoor in the west. This is why Ms. B has been advised to move from her home. However, while the candidate destinations are further upstream, both are still along the same river course, which is why she cannot agree to move to either of the destinations offered. Staying where she is and continuing to depend on groundwater is her only remaining survival strategy.

Lower reaches of the Ejene: The administrative district of Cehe lies in the lower reaches of the Ejene River. The district has Government approval to engage in cross-border trade with Mongolia. For two weeks, from June 4, 2002, the border here was opened and, in response to the need for accommodation and restaurant facilities for the many people who gather in this northern border area to work in service jobs, many inhabitants from the surrounding areas were attracted to the area during this period. To the south of the area lies the ruins of an old temple, popularly known as Laodongmiao, where there is evidence of an abandoned airport – the remains of the special military agency built by the Japanese. Further south is an abandoned settlement called Sogonoor sum, named after the dried up lake, built during the People's Commune Era. Formerly, cuckoos, great horned owls and common owls were found here and wolves from Mongolia have also moved into the area. In China the trapping of wolves is prohibited, so this side of the border has become a paradise of sorts for wolves. It is as if they were aware of the prohibition.

Ms. C lives within the administrative district of Cehe in the lower reaches of the area where the Ejene and Narin Rivers become braided. The house in which she has lived for 30 years is relatively close to one of these streams. She was previously a teacher and currently keeps approximately 200 head of livestock, mainly goats. However, in the spring of this year, only 10 kids were born. While her 20 camels have been left to wander the Gobi desert, they manage to find their way home when they need water. Around the stream there is a belt of agricultural land where Ms. C grows a modest amount of feed crops. While only a few poplars can seen here, the most visible plants are tamarisks (a deciduous shrub of the family Tamaricaceae; in Chinese *hongliu*; in Mongolian *sohai*; genus *Tamarix*). Ms. C has been told by a local official to move to a settlement in Sogonoor sum, where she would be provided with a new residence. However, Ms. C says she does not want to move.

This example raises the question of why it is necessary to pressure people to move. If Ms. C were to move she would not be able to sustain the kind of livelihood she currently does. The majority of people claim that even if, for example, she was provided with a dry feedlot facility, she would not be able to maintain her 200 head of livestock. In addition, it is unlikely that she would be provided with an alternate form employment. It seems to me that as long as Ms. C is healthy and can continue her current lifestyle, she will have no desire to move away.

Conclusion

Many researchers are participating in the "Oasis Project" and in doing so have visited numerous locations around the Heihe River valley. In the course of their work they have managed to capture the experiences of people in the grip of the great uncertainty associated with the policy of "ecological migration." For the purpose of conducting serious comparative investigations of case studies, an international symposium was convened in Beijing (Alateng & Dayu, 2004). As an additional initiative, this book was published to provide a comprehensive overview of the current state of environmental policy in China.

It should be noted that none of the three case studies presented in this paper are examples of "ecological migration" post implementation. Rather, they describe the conditions on the "eve" of the policy being implemented.

In our capacity as contemporaries of the people affected by "ecological migration", and as fellow human beings, by observing and analyzing the implementation of this policy, and through offering positive suggestions based solid empirical research, we need to ask ourselves how we as researchers can better contribute to solving the global environmental problems that confront us. I hope that in this sense too, we are on the "eve" of a new era.

References

Alateng & Dayu, 2004, Shengtai-yimin guoji taolunhui zai jing juxing, *Guangxi Minzu Xueyuan Xuebao (Zhexue Shehui Kexue Ban)*, Di wu qi, p. 149. {Chinese}
 (Alateng & Dayu,2004, Beijing international symposium of ecological migration, *Journal of Guangxi University for Nationalities; Philosophy and Social Science*, vol. 5, p. 149.)
Chugok Shakaikagakuin Minzokugaku Kenkyujo. Nihon Sougou Chikyu Kankyogaku Kenkyujo, 2004, *Kokusai Shinpojiumu 'Seitaiimin – Jissen to Keiken' (Shoroku) 7 gatu 30 nichi- 31 nichi*, Pekin Chugoku Shakaikagakuin. {Japanese}
 (The Institute of Ethnology and Anthropology, Chinese Academy of Social Science & Research Institute of Humanity and Nature, 2004, *'International Symposium – Ecological Immigration: Practice and Experience' Abstract*, at the Chinese Academy of Social Science, Beijing, 30–31 July 2003.)
Dong zhengjun, 1952, *Juyanhai (Ejina-Qi)*, Zhonghua Shuju Chuban. {Chinese}
 (Dong zhengjun, 1952, *Juyanhai (Ejene-Banner)*, China Publishing Company.)
Ejina-Qi Zhi Bianji Weiyuanhui, 1998, *Ejina-Qi Zhi*, Fangzhi Chubanshe. {Chinese}
 (Editorial Board of Ejene Banner Annals, 1998, *Ejene Banner Annals*, Fangzhi Publishing.)
Momiyama Akira, 1999, *Kan-teikoku to Henkyou Syakai – Cyoujyou no Fukei*, Chukoushinsho. {Japanese}
 (Momiyama Akira, 1999, *The Han Dynasty and Local Society – A View of Chang Cheng*, Chukoushinsho.)
Seyin, 2003, *Juyan Gudi – Heihe Liuyu de Renwen Shengtai*, Sichuan Renmin Chubanshe. {Chinese}
 (Seyin, 2003, *Juyan as Home – Human Ecology of Heihe basin*, Sichuan People's Press.)
Sven Hedin, 1931, *Across the Gobi Desert*. London, Routledge (translated from the German by H.J. Cant).

The groundwater resource crisis caused by "ecological migration"

Case studies of Mongolian pastoralists in Ejene Banner, Alasha League in the Inner Mongolia Autonomous Region

KANAKO KODAMA

Introduction

I visited Ejene *(Ejina)* Banner for the first time in the summer of 2003, just as the policy of "ecological migration" was being fully implemented. A "banner" refers to an administrative division used in Inner Mongolia Autonomous Region within China. Ejene Banner is located in the far western part of the region, approximately 600 km west of Bayanhot *(bayanhaote)*, which is the center of Alasha (Alashan) League. In the 600-kilometer expanse between these two centers lies the Gobi Desert (known in Mongolian as *gobi*); a striking sight after traveling through this area is the sight of the wooded riversides dominated by the deciduous poplar trees, called *huyang* in Chinese (*Populus euphratica*; Salicaceae; in Mongolian *toorai*). This is Ejene Banner, a historically important station on the Silk Road, referred to previously as the "corridor of poplars." Although average annual precipitation here is only 39 mm, Ejene Banner is located at the lower reaches of the Heihe River, China's second *inland* largest river, which flows down from its source in the Qilian Mountains. While the area can truly be referred to as an oasis in the desert, this region containing a stretch of vivid green riverine woodland is regarded as being one of the sources of sand storms and yellow sand. By extension, "poplars" have become the byword for environmental conservation. Yellow sand is currently responsible for causing ever-increasing and serious damage in areas as distant as the Americas, and it was in order to prevent sand storms and yellow sand, and to protect these poplar forests, that the policy of "ecological migration" was implemented.

The aim of the "ecological migration" policy is to encourage pastoralists that live within the poplar forests – the predominant riparian vegetation –

to move out of these wooded areas. The pastoralists mean the people who engage in grazing their livestock in the river woodlands and the Gobi desert. In this area, the policy affects a total of 1,500 people and 100,00 thousand head of livestock (Table 2-1). The number of migrants produced by the policy corresponds to approximately 10 percent of Ejene Banner's total population and the period of implementation was the three-year period from 2001 to 2003.

The considerable ramifications and scope of this project have prompted the question of why such a policy, which could be viewed as a form forced migration, needs to be implemented in Ejene Banner? In addition, how would the pastoralists who have lived in the poplar forests for a considerable period be able to sustain themselves after they have been resettled? Will affecting the ecological migration policy really contribute positively toward conservation of the environment? It was with these questions in mind that I undertook field surveys in Ejene Banner in 2003 and 2004.

The "ecological migration" policy in Ejene Banner

What is the "ecological migration" policy?

Before we discuss the ramifications of the policy, a more detailed outline of the "ecological migration" policy is necessary.

The "ecological migration" policy was designed to protect poplar forests by relocating pastoralists and their livestock to leave the groves. Once the pastoralists have vacated the area, fencing is erected and grazing is prohibited within the woodlands.

According to officials of the Ejene Banner administration, there are two alternate destinations for the pastoralists and their livestock that have been forced to leave the poplar groves. One option is to relocate themselves at newly established concentrated settlements in the outskirts of the center of Ejene Banner (seat of the banner government) or in several existing villages. The other option is to go to Mazongshan sum (village) located more than 300 km from the center of Ejene (Fig. 2-1). Both have been reported that the pastoralists will be provided with a fixed abode with a shed for their livestock. People who opt to migrate to Mazongshan reigon can choose a house in the migrant village of outskirts of the center of Ejene Banner or in the Mazongshan sum.

Families that move to migrant villages under the "ecological migration" policy will inevitably face major lifestyle changes. Specifically, they will have to raise their livestock in sheds and find employment or start a business.

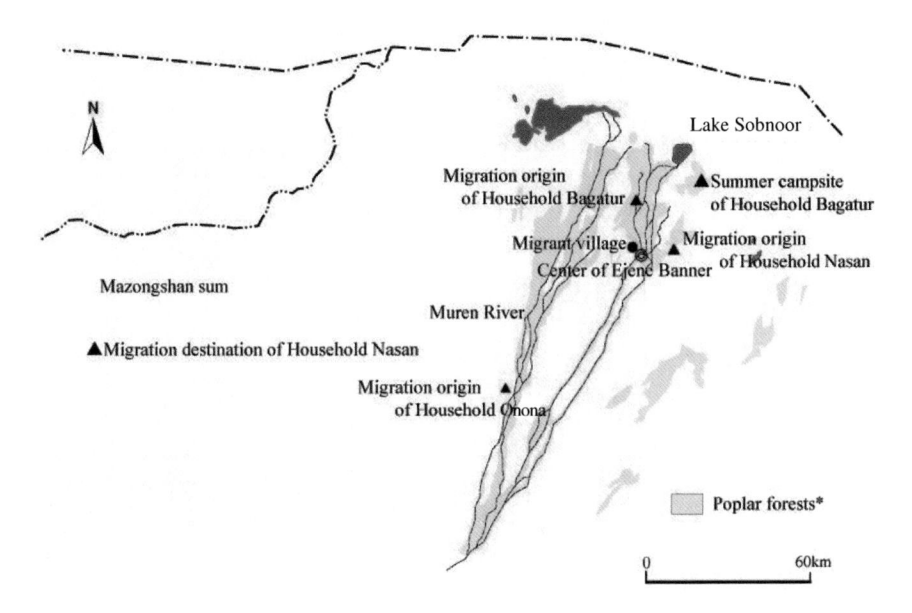

Fig. 2-1. Location of families included in this "ecological migration" case study (lower reaches of the Heihe River in Ejene Banner)

Note: Based on the map, "Waterscape Map of the Heihe River Basin, China" (1998).

* Areas marked as "poplar forests" on this map correspond to "oasis farmland" and "forest land" on the original map.

Table 2-1. Details and targets of ecological migration policy

Construction items	2001	2002	2003	Total
Ecological migrants (persons)	500	500	500	1500
Livestock (head)				100,000
Fenced poplar forests (ha)	6667	6667	6667	20,00
Fields for fodder (ha)	667	1080	920	2667
Electric pumping wells (units)	30	40	40	110

Note: Based on data collected during field surveys.

Table 2-2. Extent of ecological migration policy implementation (no. households)

Migration destination	2002	2003	2004	Total
Migrant villages on the outskirts of the center of Ejene Banner	146	14	235	395
Migrant villages constructed in villages	0	0	94	94
Mazongshan sum	19	0	0	19
Total	165	14	329	508

Note: As of December 2004
Note: Based on aural surveys of Ejene Banner officials.

Raising livestock using sheds will mean that, instead of grazing livestock on grasslands, they will be housed in stalls within the migrant villages and fed on fodder produced specifically for livestock. Migrants will be required to grow their own livestock feed, and consequently, will be provided with free access to land and access to a well with an electric pump (Table 2-1). The total area to be dedicated to feed crops is 2,667 hectares, which is approximately that of the total amount of agricultural land dedicated to cultivation in Ejene Banner in 2001. Given that irrigation is indispensable for cultivating feed crops, 110 electric pump wells have been erected in the area (Table 2-1). The people who opt for resettlement in a migrant village have another way to adopt new livelihoods – that is, they will have to switch from grazing to working in secondary or tertiary industries such as transport, commerce or the food industry. However, those migrants who go to Mazongshan sum will not be required to change their form of livelihood and will be engaged in the grazing of camels and goats.

Conditions of ecological migration at the time of implementation

According to the administrative officials of Ejene Banner, 508 families will have participated in the ecological migration project once it is completed (Table 2-2). Of these, 395 families will have been relocated to migrant villages on the outskirts of the center of Ejene Banner, while 94 will be moved into the "migrant village" constructed at the center of the village. The number of families being relocated to Mazongshan sum is 19.

According to the original plan, the ecological migration policy was scheduled for completion in 2003, but according to Ejene Banner officials the project was extended to 2004. While the assignment of families to migration destinations had already been concluded by 2004, a total of 91 houses still had to be built in the migrant villages in December 2004. In addition, no migrants have been moved to Mazongshan sum since 2003.

Thus, while the number of families that were successfully relocated to migrant villages by the end of 2004 was supposed to be 398 – the total less the 91 families for which houses had not yet been constructed and the 19 families that had migrated to Mazongshan sum – the migrant villages were quiet in both 2003 and 2004. This is why, as was suggested in Chapter 1, this entire process could only be referred to as representing the "eve of ecological migration" as too many families have opposed the policy since it was initially promulgated, or have hesitated to accept it. There are also families in which only some of the members have migrated and yet others that spend most of the year in their old grazing lands. One pastoralist, a woman in her 60s, has not even seen the house assigned to her in the designated migrant village. Ejene Banner administration officials understand the situation very well. At the end of 2004, they told me that the only people living in the migrant villages were children attending school and their grandparents, who look after them.

Why have the pastoralists been so hesitant to embrace the concept of "ecological migration"? In the next section I will report on the current status and problems affecting the implementation of the ecological migration policy.

The current status and problems associated with ecological migration

Migration to Mazongshan sum – Household Nasan

The first people to migrate – soon after the ecological migration policy was launched – were the camel feeders who lived in the poplar woods. These people moved to Mazongshan sum rather than a migrant village, and in 2003 I decided to pay them a visit. It was a long journey. Traveling westward through the vastness of the Gobi desert from the center of Ejene Banner, after some 320 kilometers and four and a half hours in a four-wheel drive vehicle, I finally arrived at Mazongshan sum. Once there, I managed to meet an ecological migrant couple in their 60s (Fig. 2-1). The husband is Mr. Nasan (a pseudonym) and the settlement is known as *maihan hara*, named after a black mountain in the area that is shaped like a tent. There is no public transport here, no electricity and mobile phones do not work.

Household Nasan were relocated to Mazongshan sum from the poplar forests in the spring of 2001. They have been provided with a house and a livestock shed in a migrant village in the outskirts of the Ejene Banner center (Fig. 2-1). Mr. Nasan made time to explain the reasons underlying their relocation.

> We were told that were would no longer be permitted to live there [due to the ecological migration policy]. Our grazing lands had become deteriorated, and both agriculture and erecting fences is hard work. We wanted to increase the number of livestock and this is why we migrated.

Household Nasan had been forced to move to the Mazongshan sum once before. That time was with 200 camels for one year between August 1975 and September 1976 during the People's Commune Era. As a result of this earlier experience of migration, he was familiar with the Mazongshan sum and this may have influenced his decision.

Mr. Nasan confided to me that at first he felt that the decision to relocate to Mazongshan sum was a mistake. Instead of increasing his herd of livestock, soon after migration he lost close to half of it. The animals that had been raised in the poplar woods were not accustomed to the Gobi desert environment, and many of them became sick or infertile. In 2002, for example, approximately 30 goats, two mature and two year-old camels died of disease. The monetary value of this loss was at least 6,000 yuan (1 yuan was approx. 0.83 $, as of Nov. 2004). The camels are still not used to the grazing land in the Mazongshan sum and continually try to make their way back to the poplar woods, more than 300 km away. Of course, each time they do this Mr. Nasan has to bring them back.

Furthermore, soon after migrating in the spring of 2001, they incurred considerable damage through a fire. All of their camel hair and cashmere, which they sell for money, as well as household items such as tents, felt and rugs, were burned. According to Mr. Nasan's wife their losses amounted to 20,000 yuan, and while the Ejene Banner government compensated them with 4,000 yuan, bedding, flour, rice and oil, this was considerably less than they lost.

The harshest aspect of life in this area is securing water. It is extremely difficult to dig for and strike sufficient water for establishing a well. In most cases, even after striking water, the water turns out to be unsuitable for human consumption. There is a two-meter open well in *maihan hara*, but it is only used for livestock as the water is not potable. Mr. Nasan has to carry drinking water from a distance of more than 50 km. He draws water once every two days if traveling by motorcycle and once a week if going by four-wheel drive vehicle.

Household Nasan's annual income in 2002 was approximately 6,400 yuan, which is approximately one third of the average annual household income (17,000 yuan according to an verbal survey) in Ejene Banner. In addition, it is exceedingly difficult harvest grass in the area, and the cultivation of feed crops is even more difficult, which means that the Nasans are forced to purchase livestock feed. In 2003, they spent approximately 1,200 yuan on feed, corresponding to approximately 20% of their total income. Since relocating, their annual income has decreased considerably.

Most importantly, the primary objective of the ecological migration policy, which was the conservation of the poplar forests, has not been achieved. According to Mr. Nasan, the poplar woods where they previously lived are currently in worse condition than before he migrated. He had dug ditches around the land he used to occupy in preparation for the fencing for enforcing the grazing ban – spending more than 500 yuan of his own money to do so – but the government has not yet erected the fencing. It also appears that some Han Chinese living on a farm nearby stole some of his old fencing to use as firewood and the now total absence of fencing around his old land has permitted livestock belonging to other households to enter the area freely and has resulted in even greater degradation than it was before they migrated.

Thus, since they migrated the Nasans have lost livestock, experienced a fire, and have generally suffered terribly. Nonetheless, the condition of the grazing land in the area has improved in the last two years and they believe that life is likely to get better. It was impressive meeting these people who, despite having lost so much and having faced innumerable difficulties, did not regret migrating. So what is it about the Mazongshan sum that is so appealing to household Nasan? The answer has to do with the quality of the vegetation for feeding their livestock and its abundance. When I asked Mr. Nasan's wife about the local vegetation, she listed approximately 10 types of local grass that camels like to eat. She told me the following:

The grass in Mazongshan is better. It's difficult to raise camels in *gol* (in the poplar woods) now. The only available feed there is *ümeihei-ebüsü* (perennial herb of the order Zygophyllaceae; in Chinese *luotuopeng*; botanical name *Peganum harmala*), *boyan* (perennial herb of the order Leguminosae; in Chinese *kudouzi*; botanical name *Sophora alopecuroides*), and *sohai* (deciduous shrub of the order Tamaricaceae; in Chinese *hongliu*; botanical name *Tamarix* spp.) People in Mazongshan sum say that the grazing land around here is getting worse, but from what I see it's much better here than in *gol*.

It's surprising to me that the poplar woods, which initially appeared to me as being a rich oasis of vivid green, are inferior to the vast bleak Gobi desert as a source of feedstock.

Despite its large size, there are less than 100 households in Mazongshan sum. The reason that people are reluctant to come here is related to water supply. Due to the limited availability of water, extensive expanses of grazing land and rich vegetation remain undeveloped. Formerly, the riverine poplar forests at the lower reaches of the Heihe River were so fertile that there was no need to utilize the Mazongshan sum for grazing. Now however, the poplar woods have degenerated to the point where migration to Mazongshan sum is necessary. Mazongshan sum is surrounded by the vastness of the Gobi desert and lies amid many mountains. When it rains the grass grows profusely. But when it doesn't rain, nothing grows here. The entire Mazongshan sum is a barren wilderness, described by Sven Hedin in his travel report as the "kingdom of death" (Hedin, 1944). In the space of the past 70 years, the poplar forests, once referred to as "an earthly paradise" (Hedin, 1931), have degenerated to a state worse than the "kingdom of death".

An official of the Ejene Banner administration told me in 2003 that from then on, anyone wishing to migrate to the Mazongshan sum would be dissuaded from doing so due to the difficult living conditions. Whether it is this dissuasion that has resulted in no families having migrated to Mazongshan sum since 2003 is not clear. Instead, ecological migrants have been encouraged to opt for a house and livestock shed in a migrant village.

An example of failed static livestock farming – Household Bagatur

Next, I will present the case of household Bagatur (pseudonym), who received widespread media coverage as being examples of the successes that were possible with livestock sheds within the context of the ecological migration policy. The couple, now in their 60s, and their families had been pastoralists in the poplar forests for many generations southeast of Lake Sobnoor ("summer campsite" in Fig. 2-1). They told me that up to the 1960s, the grass grew as high as a camel, but that it stopped flowing in the 1970s that a 10-meter deep open well that supplied enough water for more than 300 camels per day dried up and many of the poplar woods withered. According to Mr. Bagatur, it had truly become *Khara Khoto*. *Khara Khoto* are the famous ruins of a city that flourished during the Shihshia and Yuan periods, but that was eventually abandoned when its water supply dried up. In modern times, the reason for the water supply drying up can be attributed to extensive irrigation that is practiced along the banks of the middle reaches of the Heihe River.

In 1991, the Bagaturs decided to move away for the winter to other grasslands, in order to graze their livestock elsewhere ("Migration origin" in Fig. 2-1) What they found was that the new grazing land was too limited in extent to sustain a large number of livestock. As a result, the livestock in their possession, which had numbered more than 300 at the onset of the ecological migration in 2002, decreased by 120 animals to approximately two thirds. The following year, in 1992, Lake Sobnoor disappeared and their original grazing lands deteriorated further, which meant that they could not return there.

It was under these circumstances that the policy of ecological migration was implemented. In 2002, household Bagatur decided to spend the winter in a migrant village on the outskirts of the Ejene Banner center to raise their livestock in a livestock shed there (Fig. 2-1). In the autumn of 2003, Mr. Bagatur explained his reasons for moving to the migrant village, as follows:

> Now we that we are old, we can't manage to graze all our animals. When it is warm, we can manage the grazing, but in winter it is too cold. As a result, it has become difficult for us to sustain a living. Preparing firewood is also difficult, but in the migrant village we can use coal. So, when the government offered us a house we moved there because we were thinking of our retirement. The site we were allotted is approximately 670 m^2 in area, including a 60-m^2 house and a livestock shed for keeping our livestock. We had heard that after moving to the migrant village we would be assigned a 2.67-hectare field, but we have not yet received that land. (As of the end of 2004, the Bagatur's had still not received the land).

Before he relocated, Mr. Bagatur thought that raising his livestock in a migrant village would make life easier and that it would be more economically viable. However, in the winter of 2004, in his third year of raising livestock in the migrant village, Mr. Bagatur began to think that it would be better to move out of the village the following year. His reasons were that raising livestock in a livestock shed in the migrant village was very laborious and very far removed from the easy retirement he might have envisaged previously. In addition, over the three years that he had been there he had been forced to reduce the number of livestock to approximately 40 head – just one third of what he had arrived with.

For approximately four months between the winter of 2002 and the spring of 2003, household Bagatur raised 120 small livestock and two head of cattle in their livestock shed in the migrant village. However, the costs involved were prohibitive and in this period of a little over four months, they spent a total of 5,760 yuan just on maintaining their livestock shed. The two largest expenses were for the transport of hay, and the purchase of five tons of feed corn. The transport costs consisted of bringing the hay that the Bagaturs' had prepared from grass cut at their old grazing lands to the migrant village. In 2003, the Bagaturs' sold 20 small livestock to cover the costs of these various expenses and in their second year, from the winter of 2003 to the spring of 2004, they reduced their count of small livestock by a third. In all, they kept 80 small livestock and four head of cattle in their livestock shed that season. Despite this reduction, however, due to insufficient feed they were only able to keep the livestock shed running for two months. Thus, for the season from winter 2004 to spring 2005, their third year in the migrant village, they planned to cut their count of small livestock by a further one half.

As was demonstrated by this example, the principal concern associated with raising livestock using livestock shed is the preparation of sufficient feed. Since there are no fields available for planting feed crops in the migrant village in the Ejene Banner center, it is not possible to prepare enough hay or other animal feed. Even if a family still has land available that was once only used for accommodation before the migration policy, and even if they able of cultivating feed crops there, then there is the expense associated with the transportation of the feed to the migrant village.

In addition, as with Mr. Nasan, conservation of the poplar forests was not achieved and according to Mr. Bagatur the land that they left is now in worse condition than it was before. Here too, in the absence of anyone to manage the grazing land, the steel wire that surrounded the land has been stolen and the livestock of other families has been able to enter the land, resulting in the land becoming badly deteriorated.

Mr. Bagatur is now so famous that there is hardly anyone in Ejene Banner who has not heard of him. However, his fame is not due the success that he has had with adapting to new conditions, but rather because of the failures he has experienced due to ecological migration. The case of household Bagatur makes it clear that the use of livestock sheds requires considerable amounts of livestock feed, and that growing ones own feed crops is essential. According to Mr. Bagatur, even if one is able to cultivate feed crops, the current size of livestock sheds in the migrant village does not permit more than approximately 100 small livestock to be kept. With this number, which is only half the average number of head of livestock per family in Ejene Banner, income is significantly reduced. It is thus likely that a switch to static livestock raising will result in a reduced income. If this is the case, the only option remaining is to try a new kind of livelihood.

Attempts at new livelihoods – Household Orona

In the winter of 2004, the last year of the ecological migration policy, I visited the house of the Orona (pseudonym) family in a migrant village on the outskirts of the Ejene Banner center. I had heard that the family had started a taxi business and restaurant (Fig. 2-1). The Orona family's house in the migrant village is distinguished by having two mobile tents (Mongol *ger*) erected in front of the house (Photo 2-1).

The Orona family consists of Mrs. Orona, her husband and their three sons. They told me that they received their house in the migrant village with an attached livestock shed in December 2001 and that they moved there in September 2002. The Oronas accepted the offer to move into the migrant village for two reasons: to help their two eldest sons forge an independent means of livelihood, and to secure a residence to permit their youngest son to attend primary school in the Ejene Banner center. At the same time, as of December 2004, the Orona family still owned 400 head of small livestock and was raising them in the grazing lands where they had lived along the Muren River (Fig. 2-1). In order to support themselves in the migrant village, the family decided to try a new means of livelihood. Mrs. Orona enthusiastically described the process.

Photo 2-1. Mrs. Orona (second from the right) and her three sons (Mrs. Orona's husband was not present as he was grazing livestock in the grasslands in the area where they lived previously.)

Over the six-month period from October 2003 to April 2004, the Orona family unsuccessfully tried to raise 60 goats in the livestock shed of their migrant village home. Their failure was due to the fact that maintaining the livestock shed was too expensive. They spent a total of 11,000 yuan: 8,720 yuan for 60 goats that they purchased specially for the livestock shed as well as the 2,300 yuan for livestock feed and transportation of hay. In addition, the price of goat meat dropped at the time they decided to sell their goats, causing them to incur a loss of 3,000 yuan. Mrs. Orona concluded that raising livestock using livestock sheds is prohibitively expensive unless a family grows its own crops to feed its livestock. As of December 2004, the family had still not received any land for cultivating the necessary feed crops.

In the event that the Oronas are given land for cultivation in 2005, they have no intention of planting feed crops or pursuing static livestock farming. Instead, they intend to cultivate cash crops such as cotton. This is because they have heard that cotton cultivation, if successful, can bring returns of 15,000 yuan per hectare, which is equivalent to the average annual income in Ejene Banner.

Subsequent to the failure of their efforts at raising livestock using livestock sheds, the Orona family purchased two taxis in 2004 and Mrs. Orona 's husband and one of his sons now work as owner-drivers. The cost of the two taxis was 26,000 yuan. To cover this expense, which is close to their total annual income, the family drew on savings put aside for the marriage of their sons. Mrs. Orona came to realize that the market for taxis was already saturated in Ejene Banner and that it would be difficult to sustain a household on the income derived from fares alone. So, in the autumn of 2004 she erected two Mongol *gur* tents in front of their home in the migrant village and began operating a restaurant serving Mongolian dishes. As she describes it, her motivation for opening the restaurant was to "sell her livestock bowl by bowl" – instead of selling their animals alive, as before. The Orona could process them themselves and thereby sell a value-added product.

However, even now, the majority of the Orona family's income is derived from the livestock they continue to graze in the grasslands where they lived before migration. The amount they earn working in the migrant village is meager, but at the same time, the cost of living and benefits from services such as electricity and coal, is high. Mrs. Orona expects that once grazing is banned where they used to live, they will earn their living, not by driving taxis or running a restaurant, but rather by cultivating cash crops such as cotton in the field that is supposed to be provided to them.

Ecological migration – A policy that causes poverty

The most critical issue for families that are required to migrate under this policy is how to secure an income after migration. In two of the cases addressed in this chapter, those of the Nasans and the Bagaturs, there was marked decrease in their respective incomes after "ecological migration." The fear of suffering from less income is one of the main reasons that pastoralists hesitate to adopt "ecological migration."

One pastoralist, a woman in her 60s, had positive perspective of ecological migration policy and said it drives out goats from the poplar woods. Previously, goats would dig up and eat poplar saplings, causing the trees to die. This woman has just moved to a migrant village. I asked her how she intended to subsist from now on. She replied as follows:

> I will raise livestock in my livestock shed. I'll do this because that's what I've been asked to do. Some people worry that they will starve or have to turn to begging, but throughout their history Mongolians have never starved or begged. I don't think anything will happen. We will get by somehow. The government will do something for us… I'm over 60, approaching 70, so as for what I'm going to do and how, I don't even think about it. Sooner or later I will die.

This woman's response is optimistic, in that she believes the government will do something for her.

Unsustainable utilization of water resources due to irrigation

How should the government respond?

There are two approaches that the government could adopt if it chose to pursue this policy further. One way is to subsidize livestock feed, while another is to allocate land to migrants for the cultivation of feed crops by clearing poplar forests. Such an approach would result in the large-scale cultivation of feed crops in the vicinity of migrant villages. Conversely, if enforced migration to the migrant villages continues and some remedial measures are not taken, migrants trying to secure an income are likely to initiate large-scale of cultivation of cash crops.

Feed crops and cash crops cannot be cultivated without irrigation. There is no doubt that extensive irrigation would consume a huge amount of water resources. Already in Ejene Banner, the decline in river flow has led to drying up of open wells up to four or five meters deep, or to a decline in water quality. If, in addition to reduced river flows, large-scale irrigation around the riverine forests were implemented, groundwater levels would inevitably decrease further. According to Mr. Bagatur who is highly familiar with the damage caused to poplars by reduced river flows, when groundwater levels fall by more than five meters the poplars wither and die. While geographically different, the northeast of China offers an ominous example. Groundwater levels have decreased by approximately 10 meters over the last 20 years due, principally, to irrigation (Matsumoto, 2002). In short, mass irrigation would destroy the poplar forests and, if this were to happen, the ecological migration policy will have failed to protect the poplars. Irrigation in arid areas also leads to the problem of salt accumulation. The policy will thus not only have failed to protect the poplar groves, it will also have failed to prevent sand storms and yellow sand, the prevention of which was the primary purpose of the policy. The policy will thus have contributed to a worsening of the problem.

It appears that grassland-grazing bans may not be the best approach to conservation of poplar forests. As the experiences of the Nasans and the Bagaturs show, unattended grazing land is as likely of being degraded as it is to being rehabilitated. There is also a very interesting study in which it demonstrated that the presence of a certain number of grazing livestock had the effect of increasing vegetation diversity in grazing lands (Fujita, 2003)

The second "ecological migration" – the policy of "grazing discontinuation for grassland restoration"

Despite the danger posed by irrigation as a likely cause of damage to the ecological environment, a second "ecological migration" policy, based on the use of irrigation, is being applied in Ejene Banner. This policy is known as "grazing discontinuation for grassland restoration". This policy aims to protect saxaul (perennial shrub of order Chenopodiaceae; in Mongolian *jag*; in Chinese *suosuo*; botanical name *Haloxylon ammondendron*) and the surrounding vegetation, by constructing steel fencing around tracts of grazing land, and subsequently, once the livestock have been removed of the enclosed area and applying a grazing ban. This "grazing discontinuation for grassland restoration" policy will affect a total land area of 73,000 hectares and a total of 220 households. Of these households, 60 intend to move to migrant villages on the outskirts of the Ejene Banner center and pursue static live stock farming there. Each of these households will reportedly be allocated 2.0 to 2.3 hectares of land for cultivating livestock feed crops under irrigation. The area targeted by the policy lies within the Gobi desert, and has been used for grazing by camel herders. Most of these pastoralists would never have come in to contact with a hoe, requiring them to use it to earn a living may be unfeasible. I was informed that after hearing about the policy one camel herder was heard to exclaim, "Are they trying to convert us from pastoralists to farmers?"

Conclusion

From the three case studies presented in this chapter, two dangers of the ecological migration policy as applied in Ejene Banner can be identified. One is that income of migrants' drops after relocation, and the other is the danger that groundwater resources will dry up.

Due to the large-scale irrigation along the middle reaches of the Heihe River, Ejene Banner's lakes have vanished, its rivers have shrunk and its poplar forests have been decimated. Policies such as "ecological migration" and "grazing discontinuation for grassland restoration" are being implemented here despite the fact that they use irrigation that relies on groundwater. If these kinds of policies are fully implemented, they will lead to the drying up of water resources in the region, which is precisely the opposite result of that originally intended. Decreased water resources would not only lead to the failure of poplar conservation efforts and would also deprive the people here of a vital resource. If the inhabitants of Ejene Banner are left without sufficient resources for survival, then they will be forced to migrate away from the region and Ejene Banner would become nothing more than ruins – like *Khara Khoto*.

Thanks to abundant water supplied by the Heihe River, Ejene Banner is blessed with great natural diversity and unique characteristics. Moving around the region, pastoralists have long utilized this diversity in a sustainable way. Their method of raising livestock by moving from place to place, without depending on cultivation, enables sustainable utilization of the water resources of the region as well as facilitating the protection of the poplar forests along the river. What is needed is an environmental conservation policy that takes advantage of regional characteristics in this way!

References

Fujita Noboru, 2003, Sogen shokubutu no seitai to yubokuchi no jizokutekiriyo, *Kagaku*, 73(5), pp. 563–569. {Japanese}
    (Fujita Noboru, 2003, Ecology of vegetation in grasslands and sustainable utilisation of nomadic area, *Science Journal KAGAKU*, Iwanami Shoten, Vol. 73 (5), pp. 563–569.)

Hedin, Sven, 1931, *Across the Gobi Desert*, London: Routledge.

Hedin, Sven, 1944, *History of the Expedition in Asia 1927–1935*. Part 3 1933–1935 (Reports from the Scientific Expedition to the North-Western Provinces of China under the leadership of Dr. Sven Hedin – The Sino-Swedish Expedition – Publication 25).

Matsumoto Satoshi, 2002, Datsuryu sekko ni yoru sekaihatu no arukari dojo kairyo to antei shokuryoseisan no jissen, *Ajia no Keizai Hatten to Kankyo Hozen Dai San Kan*, Keiogijyuku Daigaku Shuppankai, pp. 1–15. {Japanese}

(Matsumoto Satoshi, 2002, Alkali soil reclamation with flue gas desulfurization gypsum and practice of increasing food production, *Asian Economic Development and Environment Conservation*, Keio University Press, pp. 1–15.)

New Round of Grassland Cultivation Accompanying "ecological migration"

From case studies of herders in Xianghuang Banner, Shilingol League, Inner Mongolia Autonomous Region

SUYE

Introduction

After experienced severe damages caused by the simultaneous flooding of the Changjiang, Songhuajiang and Nenjiang Rivers in 1998, and more frequent occurrences of sand storm in the late 1990s and the early 21st Century, Chinese government serious efforts to tackle the country's environmental conservation problems. The policy of "cultivation discontinuation for forest/grassland restoration" was officially launched in 2000 for agricultural and semi-agricultural/semi-grazing regions throughout China, while the policy of "grazing discontinuation for grassland restoration" has been comprehensively implemented in stock farming region since 2003. At the same time, to facilitate the transition from grassland grazing to so-called "ecological grazing"[1] in Inner Mongolia, an important grazing area within China, herders have been encouraged to adopt feedlot or semi-feedlot

---

1 In June 2002, the government of Inner Mongolia published, "An opinion of changes in production management methods and development of ecological livestock raising." According to this "opinion," the basic concept of "ecological livestock raising" is to achieve sustained development of livestock cultivation based on the protection and rehabilitation of the ecological environment. This involves initiatives arising from China's West Development Project, starting with changes to production management methods. The future development of the livestock-raising industry should progress according to the following principles: (1) In terms of raising methods: a shift from free grazing to feedlot raising or semi-feedlot raising; (2) In terms of management: a shift from extensive management to intensive management; (3) In terms of expanding the industry: place more emphasis on the quality rather than quantity of livestock; and (4) In terms of market development: look to expand from regional to national markets, and eventually into the international market.

methods for raising livestock. Shilingol Banner proposed a strategy referred to as the "enclose and migrate strategy" that can effectively be summarized using the three catchphrases of, "enclose and prohibit grazing", "down size and migration" and "intensive management." More specifically, the strategy refers to the following measures: (1) prohibit grazing, suspend grazing during spring, and practice rotational grazing in areas traditionally used for grazing. In addition, measures directed at the discontinuation of cultivation for forest/grassland restoration in agricultural lands were also proposed; (2) Control population growth in agricultural villages and encouraging people to adopt new livelihoods by seeking work in secondary or tertiary industries in local towns; and (3) Shifts in grazing land utilization from overgrazing to more efficient management; transform stock farming from natural grazing to feedlot raising or semi-feedlot raising; transform large-scale cultivation to intensive farming. The proposals are also directed at improving industrial infrastructure and management methods for agricultural production.

While the primary objective of the "enclose and migrate strategy" is to restore the system of the region, certain economic and social goals are also expected. One result of this policy, however, is the emergence of ecological migrants in some areas and the concomitant development of a variety of social problems. Ecological migration completely changes the lives of the herders who are forced to migrate, resulting in a powerful shock both to the herders and the Inner Mongolian stock farming community. This report presents case studies of ecological migration in Xianghuang (Huveetushar in Mongolian) Banner, Shilingol League, Inner Mongolia Autonomous Region, and examines the environmental impact of the new livelihoods that have been adopted by ecological migrants.

Outline of ecological migration in Xianghuang Banner

Outline of Xianghuang Banner

Forty-five percent of the Shilingol grasslands consist of typical temperate grasslands and most of Xianghuang Banner lies within this grassland environment (see Fig. 3-1). The annual precipitation in the banner is 267 mm, of which 65.2 percent falls between June and August; it is common for many months to pass without any rain. The probability of zero rainfall is 43.8 percent in spring, 28.1 percent in summer, and 43.8 percent in autumn. The only surface water consist of seasonal rivers that appear after periods of heavy rainfall. The extremely uneven temporal distribution of rainfall is an important climatic characteristic of the Xianghuang Banner area.

Xianghuang Banner occupies a total area of 496,000 ha in the southern part of Shilingol League. The banner is bordered by cultivated land which used to be pasture land of herders in Xianghuang Banner. During the Qing Dynasty the area of the banner was known as "the royal pasture." As of 2003, the average area of grazing land per capita was 30 ha, which is close to the average for Shilingol League. Xianghuang is a livestock-raising banner and is under the pressure of over grazing and a high population density. The aggregate production from livestock grazing in Xianghuang represents 56.6 percent of the banner's 2003 GDP, with 56 percent of the total population of 28,365 engaged in stock farming. Ethnic Mongolians constitute 63 percent of the banner's population, thus constituting the third-largest Mongolian population in the eight stock farming banners of the Shilingol League.

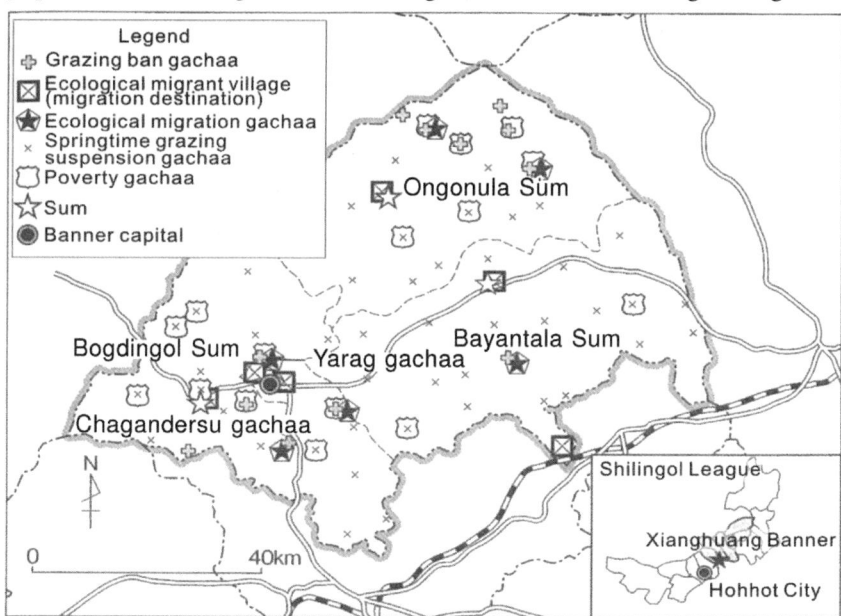

Fig. 3-1. State of implementation of the "grazing discontinuation for grassland restoration" policy in Xianghuang Banner.

Note: Based on material collected during field surveys

The main causes of damage to grasslands in Inner Mongolia are the expansion of grassland cultivation, deterioration of grasslands, overgrazing, overpopulation, and insufficient water resources (Eerdunbuhe, 2002; Suye, 2003). These factors have also contributed toward preventing the stock farming development in Inner Mongolia. Given that Xianghuang Banner is affected by all of these issues, it represents a microcosm of the entire stock farming community. Since China first launched its "reform and opening up" policy, Xianghuang Banner has served as a testing ground for various policies and systems that have been directed toward raising livestock. From 1979 to 1985, the "Yicaodingxu system" (regulating the scale of stock farming based on grass production) was implemented in Xianghuang Banner, the private ownership of livestock was introduced in 1984. In 1988, Xianghuang Banner was nominated as China's only "stock farming reform trial district," and by 1989 a system for allocating grazing rights of the banner's grasslands had been introduced. 81.1 percent of the grassland was allocated to groups of herds men households. These groups were divided into units called *hot* (a natural settlement that usually consists of approximately eight households). A contract system of land use was subsequently introduced centered on household-scale operations. However, it was not until 1996 that the boundaries between the grazing lands of each herding household were officially defined. Herders were also promised at this time some of the herders continued to use the allocated grazing land indefinitely.

In 1998, the banner was designated as a "National Ecosystem Construction Project," and in 2000 it was nominated a high-priority district forming part of a project for the prevention of sand storms in Beijing and Tianjin. Furthermore, measures such as grazing bans, grazing suspensions, rotational grazing, and ecological migration have been actively implemented as part of the "enclose and migrate strategy" in Xianghuang Banner.

State of implementation of ecological migration

Xianghuang Banner is divided into 60 *gachaas*[2]. Since 2002, grazing bans have been introduced in 12 gachaas, with the suspension of springtime grazing in effect in all gachaas. Grazing bans now cover a total area of

---

2   A *gachaa* is an administrative unit, equivalent to a "village" in China. This division is used to denote stock farming areas within Inner Mongolia.

120,000 ha, corresponding to 24 percent of usable grazing land in Xianghuang Banner. These bans will remain in effect for five years from April 1, 2003 to April 1, 2008. Herders living in the areas where grazing is banned have had to be removed from this area. While the herders are being relocated from the grazing lands in phases, eventually all of the herders will have to be relocated. In addition to providing 25,000 yuan as an "ecological migration aid" to each displaced household, the national government is offering to provide livestock feed for the first five years following migration. In principle, the ecological migrants enter into a contract with the national government to the effect that once they have received their grant they must dispose of all their livestock. Although the migrants are permitted to cut grass from the areas subject to the grazing ban once a year, the ban on grazing of livestock in these areas is permanent. Thus, while the ecological migrants have been allocated the right to use the grazing lands, they have been deprived of the right to use these lands for grazing.

The "Xianghuang Banner Enclose and Migrate Strategy Office" which was established as the organization responsible for the implementation of ecological migration, plays a central role in the planning and implementation of ecological migration throughout the banner in cooperation with other relevant departments, including the banner administration and the banner's grassland monitoring office, livestock bureau, forestry and water bureau, and grazing land management bureau. According to Xianghuang Banner's Ecological Migration Plan, as approved by the autonomous regional government, the quota for ecological migrants for the periods 2003 to 2004 was 440 households comprising 2,200 people, or 14 percent of the herders households in the banner. By August 2004, 250 households comprising 1,250 people had migrated; 64 percent were forced to move under government initiatives, while the remaining 36 percent moved voluntarily. Approximately 67 percent of the ecological migrants who have already been relocated are engaged in feedlot raising of dairy cows. A further 6 percent are engaged in feedlot raising of livestock for meat, while the remaining 27 percent have adopted livelihoods that are completely unrelated to stock farming, such as migrant labor, artisans, operating or working in restaurants, and trading.

Photo 3-1. House and livestock shed abandoned in the grasslands by ecological migrants

A sign reading "grazing ban household", as shown at the top left of the photo, is erected outside the homes of herders who have been designated as being ecological migrants. This photo was taken during a field survey in Xianghuang Banner, Shilingol League, during August 2003.

In recent years, two major dairy processing companies based in Inner Mongolia (YiLi CO. LTD and Meng Niu CO. LTD) have experienced rapid growth and became wellknown companies in the dairy industry in China. With the growth, an increasing number of dairy operations located on the outskirts of Inner Mongolia's cities and within its agricultural regions have begun to supply these two companies with milk. Private investors associated with these large dairy concerns have constructed milking facilities in these areas in order to supply milk collected from these areas to the two companies. Up until the summer of 2004, these dairy-processing companies had no milking facilities or milk collection networks in Xianghuang Banner. The herders who were raising dairy cows in the banner had to milk their cows by hands, and could only sell their milk to residents of the local town. In response to these developments and the associated increase in demand, a total of 72 ecological migrant households decided to relocate to the outskirts of Huhhot to raise their dairy livestock, which is about 500 km away from their home.

Changes in the livelihood of herders resulting from ecological migration

Between 2003 and 2004, the author conducted field surveys of ecological migration, focusing on the area of Xianghuang Banner in the Shilingol League The surveys considered two migrant villages: one located to the east of the banner's capital that consisted of 20 households and was relocated from Chagandersu gachaa; another on the outskirts of the city of Huhhot which is home to 17 households from Yarag gachaa that relocated at the same time. Grazing was banned in these two gachaas because they are located close to the banner capital, where deterioration of grassland is highly concerned. Ecological migrants from these two gachaas were the first in the banner to change their livelihood to raising dairy cows. Migration from the gachaas began in May 2002.

Based on aural surveys of four households from Chaganderesu gachaa and five from Yarag gachaa, Figure 3-2 presents a summary of gross annual household income before migration in 2001, and then again in 2002 following migration. These surveys reveal that the average income of these nine ecological migrant households increased slightly after migration, from 22,168 to 23,458 yuan. In 2001, prior to migration, sheep and cattle accounted for 77 and 18 percent of income, respectively, with only 5 percent from other sources. This remaining 5 percent included earnings from employment in non-stock farming activities, trading, and wages. These figures demonstrate clearly that up to the time of migration, the principal occupation of these people was stock farming.

The breakdown of the ecological migrants' income after relocation shows that 55 percent of income was derived from milk, 24 percent from sheep, and 17 percent from other sources. The striking points here are the shift in the principal source of income to the sale of milk and the increase in income from other sources (Fig. 3-2). It is clear from these figures that the ecological migrants cannot make ends meet from dairy farming alone, and that a large proportion (41 percent) of their total household income is generated by secondary and peripheral activities. It should be noted that the fact that the ecological migrants are selling sheep does not mean that they themselves are raising the animals. According to aural surveys, most of the households kept approximately 60 sheep with relatives or friends, effectively contracting out the grazing of the animals to these people. In addition to being an important source of income, sheep enable the ecological migrants to obtain money or meat whenever the need arises. In this way, sheep play an extremely valuable role in the lives of these people.

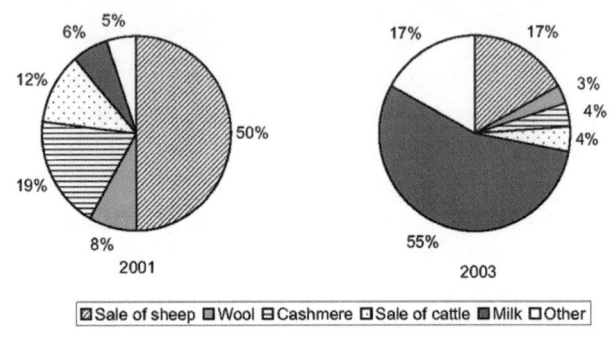

Fig. 3-2. Breakdown of annual income per household of ecological migrants

Note: Based on aural surveys of 9 ecological migrant households in Chagandersu gachaa and Yarag gachaa, Xianghuang Banner

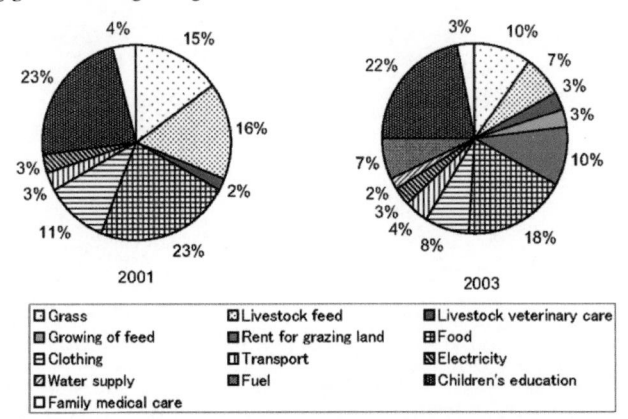

Fig. 3-3. Breakdown of annual spending per household of ecological migrants

Note: Based on aural surveys of four ecological migrant households in Chagandersu gachaa, Xianghuang Banner.

Figure 3-3 summarizes the spending patterns of ecological migrants based on aural surveys of four households from Chagandersu gachaa. Despite receiving a subsidy for livestock feed from the national government, the annual spending of the migrants was 27 percent higher after migration.

Annual spending per household increased from 11,728 Yuan before migration to 14,865 yuan after migration; at the same time, the migrants have to deal with a number of new expenditures. The new items included the cost of growing feed, rent for grazing land, water supply, and fuel charges. Upon their relocation, each family also spent approximately 10,000 yuan on home improvement, but this expense is not reflected in Fig. 3-3, neither is the approximately 7,000 yuan worth of livestock feed (based on an average price for feed corn in 2003 of 0.84 yuan/kg) received per household. If the costs of home improvement and the value of the subsidized feed were included as expenses, then households would have incurred debt of more than 8,000 yuan. The annual rate of spending is likely to rise sharply when the government feed subsidies expire in just over two years.

Another factor that influences the expenditure of ecological migrants is the increasing cost of animal feed. Table 3-1 shows changes in the price of feed over the past two years. In the period 2002 to 2004, the price of feed increased markedly, especially the price of concentrated feed that cannot be produced in the grasslands. The price of concentrated feed rose by between 18 and 70 percent over this time. With natural grazing, the cost of production associated with conventional grassland livestock raising is low. Conversely, concentrated feed is a necessity for raising dairy cattle, and the degree of dependency on outside supplies and production costs tend to be considerably higher than those used for natural grazing. As shown in Table 3-2, the feed cost for dairy cows is 11 to 20.6 times greater than it is for sheep, and 5.5 to 8.2 times greater than that for conventional cattle raising. The key to cutting expenditure for ecological migrants is to reduce the cost of feed by securing roughage rather than concentrated feed. As one solution, migrants chose to plant feed crops on their own.

Table 3-1. Changes in dairy cattle feed prices in Inner Mongolia, 2002–2004

		2003	2004	Rate of increase (%)
Concentrated feed	Feed corn	0.76–0.92	0.90–1.78	30
	Hemp cakes	1.40	2.00	43
	Bean cakes	2.00	3.40	70
	Formula feed	1.74	2.06	18
Coarse feed	Hay	0.60	0.70	17
	Silage	0.16	0.20	25

Note: Based on aural surveys conducted during field trips (to August 2004).
Units: yuan/kg

Table 3-2. Consumption and annual cost of livestock feed (2003)

	Feed consumption per head per day			Annual cost per head of livestock	
	Natural fresh grass/silage	Hay	Concentrated feed	Annual cost (yuan)	Natural feeding (yuan)
Dairy cows	15–20	4.5	4–8	4827	4827
Cattle	10–15	5–7.5		584–876	144–216
Sheep/goats	4	2		234–438	58–108

Note: Livestock feed consumption figures are based on data collected during field surveys; feed costs are based on the figures in Table 3-1.

As in the case for Xianghuang Banner, the government is restricting access to grasslands by herders under the pretext of ecosystem protection and through promoting the stock farming using feedlots. Not only have the herders received inadequate compensation from the government for their relocation, but also have to give up most of their assets and their livelihoods. In doing so, they have suffered an enormous loss and have become victims of the policy. Nonetheless, the priority is given to the protection of grasslands and only these efforts that further the purported aims of the policy are proceeded. No concerns regarding the relationship between victims and perpetrators have been raised.

Environmental impact of the new livelihoods of ecological migrants

Increased environmental burden on migrant destinations

20 families from Chagandersu gachaa settled into a needy built migrant village east of the capital of Xianghuang Banner. As of August 2004, 64 dairy cows were being raised in this village. When the author visited the village, the cowshed contained only a few cows in need of special care as the other cows were nowhere to be seen. When asked, the villagers said that the other cows were being grazed in the surrounding grasslands. The migrants were relocated here originally to engage in feedlot raising of dairy cattle as grazing is not permitted. Open grazing is considered an offense

referred to as "grazing theft" and if the migrants are caught they will be subjected to fines. Even so, the migrants are grazing their cattle in areas around the banner capital where grazing is prohibited, as well as in grasslands of other gachaas where grazing bans are also in effect. In the course of further surveys, the author learned that "grazing theft" is prevalent, both among ecological migrants and non-migrants. For example, such kind of "grazing theft" is reported being conducted by non-migrant herders in the grassland left behind by ecological migrants.

At present, six ecological migration villages (migration destinations for ecological migrants) have been established in Xianghuang Banner, of which three are already occupied. These villages are expected to be fully occupied by ecological migrants by 2005. The largest migrant village, constructed to the west of the banner capital, can accommodate 100 households, and is expected to be occupied by 2005. One reason for the desertification of grasslands to date has been addressed by the "Shilingol League Livestock Raising Journal": The increasing populations of towns accelerated the destruction of natural vegetation surrounding the towns, and led to desertification centered around the towns." (Qi, 2002a: 55). The current trend of increasing population and livestock is placing additional burdens on the ecosystem surrounding the local towns.

New era of grassland cultivation as a result of "ecological migration"

The past 100 years has been a period of large-scale grasslands cultivation in Inner Mongolia. In Shilingol League, for example, the total area of land under cultivation in 1932 was 108,000 ha, peaking in the early 1960s to more than 300,000 ha (Qi, 2002d: 926). After this time, the area under cultivation declined to a low of approximately 260,000 ha in the early 1980s before increasing again to some 300,000 ha in the 1990s (Shilingol League, Economic Statistics Yearbook). Due to a sudden rise in the frequency of sand storms in the late 1990s, policy initiatives such as "cultivation discontinuation for forest/grassland restoration" were launched in the early 21st Century. Consequently, the area of land that is under cultivation has decreased steadily, mainly due to the retirement of agricultural areas and the establishment of semi-agricultural/semi-grazing areas.

As described earlier, the concept of an "ecological stock farming industry" that satisfies the demands of both the market and the needs of environmental conservation is currently being promoted in Inner Mongolia, while the "enclose and migrant strategy" is currently being applied in Shilingol League. Associated with these efforts feedlot-farming has been introduced as an alternative to the natural grazing. As an idealized vision of this concept, dairy farms are emerging in grassland areas and many ecological migrants are choosing to work in this growing industry. As these dairy farms raise their livestock in feedlots, the infrastructural facilities associated with the farms has been justified as a grassland conservation measure, and the need to grow feed crops for these dairy feedlots has led to a new round of cultivation of grasslands. This feed-crop land (mostly corn fields) is regarded as "man-made grassland" and is considered to be an achievement of the government.

In Xianghuang Banner, ecological migrants engaged in dairy farming are permitted to grow 1.33 ha of feed crops, while general herders are permitted to grow 0.67 ha; however, an aural survey of migrants from Chaganderesu gachaa revealed an average area of feed crop cultivation per household of 1.53 ha, of which 0.87 ha were irrigated and 0.67 ha was not. This figure exceeds the government-imposed limit on the area of feed crop cultivation allowed by ecological migrants. Some of these herding households manage to increase their yield by cultivating up to 4 ha of irrigated feed crops on land leased from other herders.

Unlike grassland cultivation in the past, this type of crop farming emphasized on irrigation facilities and is considered to be a suitable method for achieving high and stable crop yields in drought-prone grassland areas. The natural productivity of grassland in Xianghuang Banner is extremely low. According to information from the Xianghuang Banner Grassland Monitoring Office, the average yield of natural fresh grass is 1,110 kg/ha. The yield is 11,250 to 22,500 kg/ha on non-irrigated cornfields and up to 45,000 to 60,000 kg/ha on irrigated cornfields. Thus the yields of non-irrigated and irrigated corn are, respectively, 10 to 20 times, and 40 to 54 times higher that that of natural grassland. For herders who have only a limited area of available grazing land, these figures are very attractive. At present there is more than 2,000 ha of irrigated land in Xianghuang Banner planted with feed crops, equivalent to 16-times the area of feed crops (included non-irrigated cultivation) grown in 1988. This is the largest area of land under irrigated cultivation in Xianghuang Banner ever.

In Xianghuang Banner, feed corn is generally planted in mid-May and harvested in late August. Thus, for the nine months between September and May of the following year, the land remains bare. This period is also characterized by dry conditions and strong winds, conditions that contribute to desertification due to wind erosion. Even an overgrazed grassland offers greater resistance to wind erosion than such bare land. It is estimated that it would take 50 to 100 years for the vegetation of the cleared grasslands of Shilingol League to be restored to its original state (Qi, 2002d: 927). The area currently cultivated for feed crops may only be a fraction of the total area of grasslands, however, the grassland is being cultivated again under questionable pretexts as it happened in the past.

The development of water resources together with feed crop cultivation

Along with the cultivation of feed crops, substantial efforts are also being invested in the development of water resources. The only way to cultivate feed crops in livestock-raising areas with no rivers, such as Xianghuang Banner, is to pump underground water. According to local people, a pumping well more than 100 meters deep has been sunk at a cost of approximately 100,000 yuan. Xianghuang Banner has had a bitter experience in the past on water resource development. According to the "Xianghuang Banner Journal" (Vol. 5, Ch. 1, pp. 243–253), from the 1970s to the early 1990s the central government, together with the regional government and local residents, managed to raise over 10 million yuan to construct 3 irrigation plants, 2 dams, 28 underground dams, and 135 wells in order to launch fish-breeding projects. Unfortunately, these facilities largely fell into disuse due to factors such as, defective construction work, careless management, severe evaporation, falling under groundwater levels, the privatization of grazing rights through the introduction of a contract system, and high operating costs. By 1992, the only facilities still in use were 35 wells.

In semi-arid areas, water is the most precious resource. Therefore, any water resource development should be undertaken with great caution. As a high rate of water consumption is a potential major factor in the destruction of grassland ecosystems, the development of sustainable water resources is an important issue and a major challenge to the region.

The necessity of ecological migration in Xianghuang Banner: the herders' perspective

An important characteristic of the grasslands of Shilingol is the frequent occurrence of droughts that arise due to irregular precipitation patterns. In 2001, the entire Shilingol League endured the worst drought in 12 years. As a result of the consequent desertification of grasslands, the number of sand storm days rose to 20 for the year, considerably higher than the usual 6 days per year experienced in the 1950s (Qi, 2002a: 55).

In Xianghuang Banner, precipitation during 2001 amounted to 200.5 mm, which is 25 percent less that average and made that year the seventh-driest year on record in the past 42 years. The drought caused a huge damage to the livestock industry. According to the banner's statistical records, in 2001 the number of cattle and sheep fell by 66 and 41 percent, respectively. Ecological migration was initiated the following year. In 2002, and in 2003 the banner enjoyed the second-highest annual precipitation (413 mm) for the previous 30 years. Abundant rainfall normally means a relatively easy year for herders; consequently, many of the ecological migrants felt that it had been unnecessary for them to migrate. Although yields of grazing grass were high in the year following migration, the migrants had left the grazing areas and were unable to take advantage of the situation. The migrants were also unhappy about the prevalence of "grazing theft" that occurred on their pasture land due to their absence. On the basis of interviews with ecological migrants, the author found that 6 out of 10 migrant households felt that migration had been unnecessary, as they believed that they could have overcome the drought by employing the *otor* method, or by reducing the number of livestock.

Conclusions

Although the phenomena of ecological migration from the grasslands and the emergence of the dairy industry are only local activities, they have caused considerable changes to both the lifestyles of herders and a restructuring of local society, as well as a new trend of grassland cultivation accompanied by the exploitation of new water resources. These changes are in the process of completely transforming longstanding traditions of livestock cultivation. With the help of modern technology and massive investment, even the issue of securing water resources, which used to be the most difficult challenge for the region, has been resolved. In the short term, this availability of water will enable high levels of production and high profits. But whether these activities are sustainable is questionable. As described in this paper, the concentration of ecological migrants in towns, the expansion of feed crop cultivation into grasslands, and the considerable pressure on water resources may lead to new environmental problems. Unlike the previous trends of grassland cultivation, the grassland cultivation is conducted on the remaining grassland and due to the heavy use of water resources, it is likely that the effects on the ecosystem will be considerably more profound than was previously the case.

The government has committed itself to providing aid to ecological migrants for five years: assistance for ecological migration implemented in 2002 will continue until 2007. Given that ecological migrants will then have to be self-sufficient, it is essential that the ecological migrants have access to cheaper feed if they are to continue raising dairy cattle. Certainly, one way for the migrants to obtain more feed is to plant more. This is likely to result in further exploitation of grassland environments that are already in a critical condition. Measures to prevent this damage to grassland areas must therefore be conceived before it is too late. The formulation of substantial aid measures and special policies for ecological migrants needs to be addressed. Furthermore, from the viewpoint of grassland protection, efforts should be made to secure livestock feed from outside of grassland rather than attempting to secure these resources from the grasslands themselves.

Acknowledgements

This paper is a summary of surveys funded by a Grant-in-Aid for Scientific Research as apart of the Junior Scientist Program (B) provided in 2003 and 2004.

References

Daoerjipalamu, 1996, *Jiyuehua Caodi Xumuye*, Zhongguo Nongye Kexue Chubanshe, pp. 87–88. {Chinese}
(Daoerjipalamu, 1996, *Intensive Livestock Raising in Grassland*, Chinese Agricultural Science Publishing, pp. 87–88.)

Du Fulin, 2004, Taikokanrinkansoseisaku no tenkai to chiikinogyo no henka, *Chiiki Chiri-Kenkyu*, 9 pp. 18–19. {Japanese}
(Du fulin, 2004, The movement of the Tuigenghuanlin (reforestation of cultivated land) policy and the change of the regional agriculture, *The Research of the Regional Geography*, 9 pp. 18–19.)

Eerdunbuhe, 2002, Caodi liyong budang yu caodi huangmohua, Eerdunbuhe deng bian, *Neimengge Caodi Huangmohua Wenti ji qi Fangzhi Duice Yanjiu*, Neimengge Daxue Chubanshe. {Chinese}
(Eerdunbuhe, 2002, The grassland desolation on account of unjust utilization of resources, *Research of Problems of the Grassland Desolation in Inner Mongolia Autonomous Region and Prevention Countermeasures against it*, Inner Mongolia University Press, pp. 152–161.)

Fukui Katsuyoshi, Tani Yasushi, 1987, *Bokuchikubunka no Genzo – Seitai, Shakai, Rekishi*, Nihon Hoso Shuppan Kyokai. {Japanese}
(Fukui Katsuyoshi, Tani Yasushi, 1987, *Aspects of the Culture of Pastoralism; Ecology, Society, and History*, Japan Broadcast Publishing.)

Peng Hougan Zhubian, 1985, *Xumuye Changyong Shuju Shouce*, Neimengge Renmin Chubanshe. {Chinese}
(Peng Hougan, 1985, *A Data Manual of Livestock Rising*, Inner Mongolia People's Press.)

Qi Boyi, 2002 a, *Shilingol-Meng Xumu Zhi*, Di yi bian, Neimengge Renmin Chubanshe. {Chinese}
(Qi Boyi, 2002 a, *Shilingol League Livestock Raising Annals*, Vol. 1, Inner Mongolia People's Press.)

Qi Boyi, 2002 b, *Shilingol-Meng Xumu Zhi*, Di er bian, Neimengge Renmin Chubanshe. {Chinese}
(Qi Boyi, 2002 b, *Shilingol League Livestock Raising Annals*, Vol. 2, Inner Mongolia People's Press.)

Qi Boyi, 2002 c, *Shilingol-Meng Xumu Zhi*, Di san bian, Neimengge Renmin Chubanshe. {Chinese}
(Qi Boyi, 2002 c, *Shilingol League Livestock Raising Annals*, Vol. 3, Inner Mongolia People's Press.)

Qi Boyi, 2002 d, *Shilingol-Meng Xumu Zhi*, Di si bian, Neimengge Renmin Chubanshe. {Chinese}
(Qi Boyi, 2002 d, *Shilingol League Livestock Raising Annals*, Vol. 4, Inner Mongolia People's Press.)

Qi Boyi, 2002 e, *Shilingol-Meng Xumu Zhi*, Di wu bian, Neimengge Renmin Chubanshe. {Chinese}

(Qi bai yi, 2002 e, *Shilingol League Livestock Raising Annals*, Vol. 5, Inner Mongolia People's Press.)
Qi Boyi, 2002 f, *Shilingol-Meng Xumu Zhi*, Di liu bian, Neimengge Renmin Chubanshe. {Chinese}
(Qi Boyi, 2002 f, *Shilingol League Livestock Raising Annals*, Vol. 6, Inner Mongolia People's Press.)
Shilingol Meng Shengtai Jianshe Bumen, 2003, *Shengtai Jianshe Weifeng Yizhuan Zhengce Fagui Bai Wen*. {Chinese}
(The Department of Ecological Construction in Shilingol League, 2003, *The 100 Regulations of Policy of Enclosure and Transfer attached to Ecological Construction*.)
Suye, 2003, Chugoku Uchimongorujichiku ni okeru sogenhakai no higai kagai kozo no bunseki, *Kankyo Shakaigaku Kenkyu*, 9 pp. 18–29. {Japanese}
(Suye, 2003, Analysis of structure of damage and mischief from grassland destruction in Inner Mongolia Autonomous Region, *Journal of Environmental Sociology*, 9 pp. 18–29.)
Xianghuang Qi Shiyan Bangongshi Bianji, 1994 a, *Xianghuang Qi Muqu Gaige Shiyanqi Wenjian Huibian*, Di yi ji. {Chinese}
(Xianghuang Banner, 1994 a, *The Collection of Documents; Livestock Raising Reform Trial District in Xianghuang Banner*, Vol. 1.)
Xianghuang Qi Shiyan Bangongshi Bianji, 1994 b, *Xianghuang Qi Muqu Gaige Shiyanqi Wenjian Huibian*, Di er ji. {Chinese}
(Xianghuang Banner, 1994 b, *The Collection of Documents; Livestock Raising Reform Trial District in Xianghuang Banner*, Vol. 2.)
Xianghuang Qi Shiyan Bangongshi Bianji, 1997, *Xianghuang Qi Muqu Gaige Shiyanqi Wenjian Huibian*, Di san ji. {Chinese}
(Xianghuang Banner, 1997, *The Collection of Documents; Livestock Raising Reform Trial District in Xianghuang Banner*, Vol. 3.)

Forest restoration without reliance upon "ecological migration": From a case study of NGO activities in Guizhou Province

YOSHIKI SEKI and XIANG HU

Introduction

This paper aims to verify whether ecological migration programs are necessary for achieving forest restoration. More specifically, we study the case of the *Tuigenghuanlin* program (Land Conversion Program from Farm to Forest), the largest forestation program ever conducted in human history, to find a way for the program to succeed without ecological migration. We selected Gusheng Village for the study, a highly sloped marginal village in Guizhou Province where 78 percent of the village's total farmland has been converted to forest in line with the *Tuigenghuanlin* program.

In a previous study, we examined the socio-economical impact of the *Tuigenghuanlin* program on the farmers' life in Gusheng Village. We found that residents of this mountainous village, where as much as 80 percent of the village farms are located on land with an inclination of 25 degrees or more and are subject to land conversion, were concerned about how their lives would be affected by the enforced changes in land use (Xiang and Seki, 2003).

Within this context, the option of ecological migration might be considered. However, there may be an alternative to achieve both the objective of reforestation and that of securing a livelihood for the villagers without relying on ecological migration. Accordingly, we have examined various measures for developing livelihoods in a mountain village. We asked a rural development NGO in Guizhou Province to initiate a project with the aim of achieving both sustainable livelihoods and afforestation via voluntary collaboration among the NGO, villagers and the local government unit.

In this paper we briefly introduce how people's lives have changed following implementation of the *Tuigenghuanlin* program, and then report on progress of the pilot project conducted by the NGO. As the project is still in progress, and unfolding on a trial-and-error basis, definitive conclusions cannot yet be made. Nonetheless, we would like to describe the prospects of this initiative based on the practical activities undertaken and also show that people can develop their lives even after land conversion from farm to forest.

Tuigenghuanlin and "ecological migration"

After the disastrous flooding along the Changjiang (Yangtse) River in 1998, the Chinese Government formulated a series of forest conservation and regeneration policies centered on the *Tianranlinbaohu* (Natural Forest Protection Program) and the *Tuigenghuanlin/huancao* (Land Conversion Program from Farm to Forest/Grassland). Ecological migration programs are encouraged under *Tianranlinbaohu* and *Tuigenghuanlin/huancao*. The *Tuigenghuanlin* program, on which this paper focuses, aims to convert agricultural land into forest in ecologically fragile areas. As a consequence of this policy, large numbers of farming households have lost their agricultural and grazing land. Ecological migration is being jointly promoted as a "package" with these forest policies.

Before discussing ecological migration as the policy instrument that followed *Tuigenghuanlin*, it is necessary to investigate changes to the lives of affected farmers following the conversion of cultivated land. We search for strategies to secure livelihoods for villagers that are compatible with reforestation.

The *Tuigenghuanlin/huancao* program has been applied mainly along the Changjiang River and Huanghe (Yellow) River basins, in steep landscapes (slopes greater than 25 degrees) where soil erosion is severe and in areas suffering from severe desertification and alkalization. The policy aims to convert agricultural land into forest or grassland to prevent soil erosion, alleviate flood damage in high-rainfall areas, and stem the progress of desertification in arid areas. While "grassland restoration" has been promoted in areas of advanced desertification, in this study we have focused on "forest restoration" in a high-rainfall area in the upper reaches of the Changjiang River.

Farmers who surrender farmland as a result of the *Tuigenghuanlin* program get compensation in the form of grain and a small amount of cash. The amount of grain provided to farmers per *mu* (1 *mu* = 0.067 ha) of land conversion is set at 150 kg per year along the Changjiang River and 100 kg per year along the Huanghe River. The cash assistance is 20 yuan per year per *mu* of land (1 yuan is approx. US$ 0.10 as of March 2005). The compensation in the form of grain and cash are provided for a period of five years in the case of planting "economic trees" (trees that have commercial value) and eight years in the case of planting "ecological trees" (trees planted for conservation). Fruit, tea and nut bearing trees are categorised as economic trees, while native species are categorised as ecological trees.

The planning targets of the *Tuigenghuanlin* program as of 2002 are ambitious. It plans to reforest/afforest 32 million hectares (15 million hectares of land conversion and 17 million hectares of planting in denuded mountains (Zhongguo kechixu fazhan linye zhanlüe yanjiu xiangmuzu, 2002: 231). The program will forest an area approximately equivalent to 85 percent of the total land area of Japan.

Article 54 of the *Tuigenghuanlin* Ordinance, which came into effect in January 2003, prescribes the promotion of ecological migration in the process of implementing the program. Land conversion from farm to forest and ecological migration can be thought of as the two wheels of a cart for the aim of ecological conservation. Retiring 15 million hectares of farmland would mean that more than 100 million farmers lose at least part of their land. Clearly, it is impossible to move 100 million people. Ecological migration can not be frequently applied without careful consideration for the purpose of reforestation.

The impact of *Tuigenghuanlin* on the lives of the people of mountain villages depends on the proportion of the total farmland to be converted under the program. Article 22 of the People's Republic of China Forest Law Implementation Ordinance, issued in 2000, states that "Steep land having a gradient greater than 25 degrees must be planted with trees or grass, and in conformity with the plans of the prevailing local governments, cultivation on such land should be gradually discontinued, and trees or grass planted instead." As a result of this law, a gradient in excess of 25 degrees is one criterion for determining whether land is subject to the *Tuigenghuanlin* program. Thus, villages with only a small proportion of steep land lose only a small amount of their farmland, while villages with a large proportion of steep land must surrender a large amount of farmland for the purpose of forestation.

If the affected area is only 20 to 30 percent of the total farmland, the effect on the village will be slight, but if 70 to 80 percent of farmland is to be converted, the effects are likely to be very severe. Even if the participants of the *Tuigenghuanlin* program can subsist for the eight years during which they will receive grain as compensation, it may be impossible for them to survive in such areas once the compensation is finished. Thus, the villagers will be faced with the option of ecological migration. In view of this scenario, our field survey focuses on a village in which 78 percent of the arable land has been retired and forested. If there is a way for the villagers to secure a sufficient livelihood in coexistence with the forest despite losing 78 percent of their farmland, migration will not be necessary.

Changes in the lives of farmers following the introduction of Tuigenghuanlin

Adaptive strategies of households in response to Tuigenghuanlin

Our research village, Gusheng Village in Qianxi County, Guizhou Province, is located at an altitude of 800 to 1,500 meters above sea level along the upper reaches of the Wu River, a tributary of the Changjiang River. There are 271 households in the village, with a total of 1,800 people. Han Chinese make up 92 percent of the population, with the remaining 8 percent comprising ethnic minorities such as the Miao. Gusheng Village is located within an area of karst topography that is characteristic of Guizhou Province and where the soil is infertile. During the Great Leap Forward *(dayuejin)*, natural forests surrounding the village were all felled. Excessive cultivation on steep land has caused severe soil erosion which exposed the limestone bedrock.

We initially conducted surveys of a random sample of 64 households in Gusheng Village on the subjects of changes in household economy following the implementation of *Tuigenghuanlin* and the attitudes of farmers to the program. The surveys were conducted over two periods: a three-week period in July and August 2002, and a two-week period in December 2002 and January 2003. The surveys found that farmers were dissatisfied with *Tuigenghuanlin*. We thus launched a pilot project in 2003, together with a local NGO, to find a way to support both forestation and people's livelihoods.

In excess of 80 percent of all the farmland in the village is steeper than 25 degrees and therefore subject to forest restoration. On average, each of the 64 households of the survey sample lost 78 percent of the land that they had previously cultivated. Given this situation, how are these farmers adapting to the conversion of 78 percent of their contract farmland to forest? We identified the following adaptive strategies available to households to maintain their livelihoods following implementation of the program:

(1) Expand livestock activities to compensate for the reduction in crop farming activities.
(2) Move to intensive farming by using the remaining 20 percent of farmland.
(3) Plant economic trees such as fruits, teas, and nuts to establish a non-timber plantation.

Photo 4-1. *Tuigenghuanlin* site in Gusheng Village, upper reaches of the Wu River

(4) Plant industrial timber species and operate as a timber plantation.
(5) Use surplus time to work as a day labor in public works projects in nearby areas.
(6) Find employment in distant urban areas (migrant labor).

Table 4-1. What kind of activity do you want to start after the *Tuigenghuanlin*? (Multiple responses allowed)

	Total (64 persons)		Under 50 yrs (36 persons)		50 yrs and over (28 persons)	
	No. of responses	%	No. of responses	%	No. of responses	%
Lovestock raising	44	68.8	24	66.7	20	71.4
Wage labor (incl. family members)	26	40.6	18	50.0	8	28.6
Forestry or fruits farming	10	15.6	4	11.1	6	21.4
No change	8	12.5	3	8.3	5	17.9
Independent enterprise	3	4.7	2	5.6	1	3.6
Total	91	142.2	51	141.7	40	142.9

If the villagers can sustain a livelihood by a combination of the above adaptive strategies, they should face no major problems. The attitude survey showed that the most favored strategy of affected farmers is to expand their livestock activities. Table 4-1 lists the responses provided by the 64 household heads to the question of what activity they intended to adopt after the Land Conversion Program took effect. Approximately 70 percent expressed their desire to expand their livestock-raising activities. The next most favored activity is wage labor outside the village.

Only 15.6 percent of household heads expressed any confidence in the prospects of managing tree or fruits plantations. Considering that this area is characterized by relatively high rainfall – a positive factor in tree growth – this is a low figure. Our survey of family finances showed clearly that even the highly favored strategies of livestock raising and working outside the village were not as lucrative as the households had expected. As long as the compensation from the government continued, conditions might be good, but the farmers were very worried about how they would manage once the eight-year compensation period finished. We now consider the relevant importance of these problems in order.

Difficulties associated with expanding the scale of livestock raising

Considering the high expectations associated with raising livestock, villagers found it very difficult to expand their operations. Table 4-2 shows the changes in the average number of livestock per household in Gusheng Village following the implementation of *Tuigenghuanlin*. In the first year of forest restoration, the average number of cattle and pigs per household (chickens excluded) declined.

Table 4-2. Changes in the average number of livestock per household (N = 64 household)

	2001: before the *Tuigenghuanlin*	2002: after the *Tuigenghuanlin*
cattles	1.13	0.77
pigs	3.12	2.48
chickens	15.5	37.0

There are two main reasons for the decrease in the number of cattle: (1) The banning of grazing inside the tree plantation; and (2) The fact that draft animals were no longer required because of the abandonment of farmland. As grazing in forests is banned by the *Tuigenghuanlin* Ordinance, it became difficult to raise livestock due to lack of grazing land. Some farmers sold off the cattle that were no longer needed as draft animals.

The number of pigs also decreased following the implementation of *Tuigenghuanlin*. This trend is explained by the reduction in feed corn production associated with the discontinuation of cultivation. Grain provided by the government as compensation was barely enough to feed the people, and there was no way to acquire enough feed to expand pig stocks. In fact, we found that the greater the number of pigs owned by a farm household, the more strongly the household felt that the grain compensation was inadequate.

In addition, in the case of both cattle and pigs, even if farmers desired to increase their stock of animals, they generally lack the capital to do so. One farmer, Mr. A (aged 32), expressed his situation as follows.

> I would like to raise more livestock, but I don't have the money or a shed to keep them in. If I could borrow some money, I would buy two or three pigs and one or two cattle. I think we can improve our lives, but repaying a loan would be difficult. It looks like I might have to go away to work in urban areas. If I could borrow some money for my livestock raising, I could go away and earn money for a while. Then, after paying back the loan, I could come back to concentrate on my livestock. I'm worried, though, about my lack of skill with livestock. If my animals died it would be a disaster. I really don't know if I could succeed or not.

As this example shows, even though farming households would like to buy livestock, they don't have the capital to do so, thereby increasing their reliance on finding jobs outside the village. The only economic activity that we found to be expanding in Gusheng Village was poultry; however, the only people involved in the expanding poultry operations were the village leaders, who were able to obtain "poverty support loans" from the government. The village leaders also lacked experience in large-scale livestock raising and technical training, and when we visited the village again in August 2004 we learned that most of their chickens had died.

Wage laborer in urban areas as a form of "voluntary ecological migration"

Significant barriers must be overcome when earning a living raising livestock following the land conversion from farm to forest. As a result, the most common adaptive strategy undertaken by farmers is to work away from home. The term for this practice, "outside work" *(dagong)*, refers to one of two situations: working as a day labor in public works projects in nearby areas while continuing to live in the village ("local day labor"), or leaving the village to live and work far away in urban areas ("urban migrant labor"). This latter strategy could be referred to as "voluntary ecological migration."

We therefore sought to determine to what extent the activities of "local day labor" and "urban migrant labor" increased following the introduction of the *Tuigenghuanlin* program. Table 4-3 presents a comparison of the average annual number of "outside labor" days per household before and after implementation of the forest restoration policy: "local day labor" did not increase at all, while "urban migrant labor" has increased.

Table 4-3. Changes in the average annual days for "outside work" (days per household) (N = 64 household)

	2001: before the *Tuigenghuanlin*	2002: after the *Tuigenghuanlin*
Local day labor	76.7	75.2
Urban migrant labor	167.3	195.8

In the case of Gusheng Village, "local day labor" employment is typically in a nearby dam construction project and road construction projects. Naturally, unless the number of projects increases, the average level of day-labor work cannot be expected to increase. Within individual households, the number of days worked did increase in some households, but in other households the number of days declined. Villagers were scrambling to compete for the limited number of work opportunities available to them.

The result is that a growing number of villagers are moving to the cities to seek work. It is expected that more villagers, especially the young generation, are going to adopt this strategy because it is difficult to adopt other strategies. The phenomenon of traveling far from home to seek employment might be categorized as "voluntary ecological migration." However, as this option depends largely on continued economic growth in the coastal cities, its future is uncertain. This strategy is only possible for the physically tough young generation. Even among this generation, there are farmers who choose to stay in the village after finding that they could not adapt to urban life. For example, Mr. B (aged 32) explains his experience as follows.

> Last year I went to work for one month in a sawmill in Guangdong Province, but the hot weather in the south caused me health problems. I couldn't continue and so returned home. I could only make enough money to cover my travel and living expenses. For two months this year I worked at a nearby construction site, but I was cheated. After a new site manager was appointed I didn't get my wages. People deceive me because I am a bit naive. I feel bitter about it! I don't want to go away to work anymore. Anyone with a low level of education, or who is naive, will either be cheated or bullied. I am a farmer, so farming is what I want to do. My children are still young, and I still have to pay school fees. It's really tough!

Of course, people who move voluntarily and enthusiastically to work in the urban areas and are able to adapt to the changed lifestyle should be encouraged, however, the degree to which a farmer can adapt to non-agricultural work and the urban environment depends on their personality. It is a harsh situation that compels people such as Mr. B to seek work as laborers in the city. The "ecological migration" that has arisen following the implementation of *Tuigenghunalin* should not force the entire village to leave. A better approach would be to ensure to as great a degree as possible that each individual is free to choose whether or not to migrate.

Reasons why farming households see no prospects in forest management

If the land conversion program was implemented ideally, we would expect that local people would see promise in the strategy of managing planted forest or fruit trees; however, as is evident from the survey results presented in Table 4-1, only a very small proportion – 15.6 percent – of farming households in Gusheng Village believed that they could make a livelihood in this way. Why is this?

In Gusheng Village, the government is taking responsibility for selecting varieties and distributing all seedlings to be used for *Tuigenghuanlin*. Three coniferous species – Diansong *(Pinus yunnanensis)*, Liushan *(Cryptomeria fortunei)*, and Dianyang *(Populus yunnanensis)* – account for 63 percent of the trees to be planted, with fruit trees accounting for less than 20 percent. The timber produced from the three species has a low market price, so it cannot be expected to generate substantial cash income; this is why farmers have no desire to plant such trees. The government's plan of *Tuigenhuanlin* is to adjust the ratio of "ecological trees" to "economic trees" to 80:20. The ecological trees, which must account for at least 80% of the trees planted at *Tuigenhuanlin* sites, are mainly coniferous species used for reforestation such as pine, cryptomeria, and poplar. Farmers, however, are more eager to plant economic trees than ecological trees; their lives will be further impoverished if the government ratio is strictly implemented. Mr. C (aged 52) explained his view on forest management as follows.

> At the moment, villagers don't really want to plant the distributed seedlings, but do so in order to receive food compensation. This is why villagers put minimal effort into managing the plantations. They are not taking care of the seedlings immediately after they are planted. They don't have much confidence in the planted trees. And even if they grow well, they don't think they will be able to make any money out of them.

Another reason to discourage farmers to manage the forests properly is that the government prohibits intercropping in the new plantations. The local government foresters think that villagers do not have enough technological capability to practice agroforestry. One provincial forester responded to our interview in January 2003 as follows:

> Because the farmers have insufficient technological capacity, large-scale intercropping might spur soil erosion. If farmers intercrop soybean, soil erosion cannot be prevented. Moreover, if farmers cultivate or harvest in *Tuigenghuanlin* sites, planted trees will be damaged. Because the level of technological skill among farmers is not necessarily high, all members would not be able to practice agroforestry techniques, even if we would give them training.

On the other hand, most of villagers complain about this government distrust toward farmers. As an example, Mr. C (aged 52) reflected on government policy as follows:

> If the government does not trust the farmers, the technology will not reach the farmers. With the preconception that the farmers cannot do something, the spreading of technology becomes difficult. I think that agroforestry is a technology that almost all farmers can use successfully, if the technical training is done properly. In the end, it is necessary to trust the farmers for the policy to succeed. After all, any splendid policy will fail if the government does not trust the farmers.

Our field study has shown that, contrary to the premises underlying *Tuigenghuanlin*, the farmers engaged in illegal agroforestry practices have properly managed the trees. In fact, the growth rate of seedlings is excellent in the illegal agroforestry lots as the farmers use compost that benefits the tree plantations (Xiang and Seki 2003).

Because of two policy obstacles, top-down distribution of seedlings and prohibition of intercropping, that discourage farmers, the only incentive for farming households to invest effort in tree planting is to superficially fulfill the survival-rate quota for the seedlings to qualify for obtaining compensation. For the farmers, the planted trees are seen as no more than a "coupon ticket" for obtaining compensation. As a result, farmers are apprehensive about what will happen after the government ceases to provide food compensation. In reality, the farmers are planting the trees reluctantly in order to receive food. Therefore, it is quite likely that some of them, when faced with the difficulty of surviving without food compensation, will again clear the land to plant crops. In response to our question of "What will you do after eight years when the food compensation is discontinued?" a remarkable 43.8 percent of households replied that "they would have no option but to clear the land again." Therefore, it is important to question whether ecological migration is the only way to avoid this land being cleared after the eight year compensation period, or whether there is a way for people to earn a secure livelihood without resorting to clearing the reforested land.

Collaboration among local people, NGOs and the government

Participatory development or endogenous development

Since the introduction of the *Tuigenghuanlin* program, mutual distrust between farmers and local government has also increased. From the viewpoint of farmers, the government is not responding adequately to their concerns regarding the security of their livelihoods. From the government's perspective, it feels that despite doing its utmost to provide assistance, the farmers fail to follow its guidance and appear to be interested only in receiving food compensation. We can thus anticipate that without any revisions of the program many farmers will be compelled to reclaim land allocated to the timber plantation for cultivation after compensation is ended (Xiang and Seki 2003).

In order to avoid this pessimistic scenario, we decided to initiate a project with the permission of the provincial government to support both the forestation and people's livelihood in Gusheng Village. We contacted a Beijing-based environmental NGO, "Friends of Nature," and a local NGO established by farmers in Guizhou Province, "Caohai Farmers Association," and asked them to become involved in the project. Given the growing distrust between the farmers and government, we felt that the involvement of third parties such as the two NGOs might help to rebuild a relationship of trust between the two sides and help to find a way to achieve both afforestation and a secure livelihood for the farmers. The project is financed by a grant-in-aid provided to "Friends of Nature" by the Japan Fund for Global Environment, as well as some of our own funds.

The Chinese government's forestation planning is colored by the traditional "mass mobilization" approach with inadequate participation of local people that makes people more reluctant to manage the forest (Hirano

2002). With the inclusion of participatory development methods, new possibilities might emerge for afforestation projects. In fact, long-established donors such as the World Bank and the Ford Foundation are undertaking various projects that aim to incorporate Western-style participatory development methods into Chinese forest management.

While we refer to participatory approach, the endogenous development method, which focuses local peculiar logic, is more valued to avoid just "participation" in the program designed by donors.

According to the Asian endogenous development *(neishengshifazhan)* theory, every region has a specific form of self-organizing development in the context of its natural environment, history, and culture. The form of self-organized society is "created spontaneously by conforming to its specific natural environment, based on its cultural assets, in accordance to its historical conditions, and with reference to foreign knowledge, technology and systems" (Tsurumi, 1996). We consider that the best way to solve the various problems facing Gusheng Village following the implementation of *Tuigenghuanlin* was to employ methods developed spontaneously within the specific natural, cultural, historical, and political setting of Guizhou Province.

Profile of the Caohai Farmers Association

The background to the formation of the Caohai Farmers Association is as follows. In 1992, while the black-necked crane was nationally designated a first class protected species, the entire Caohai Wetlands in Guizhou Province were designated as a national nature reserve. Fishing and hunting, which until then had supported the livelihood of local people, were restricted, and reclaimed land was restored to its former wetland status. As the local people around the wetlands could no longer earn a living, ecological migration was planned.

Some people who did not want to leave the area refused to emigrate, instead striving to find a way to successfully marry ecological conservation with economic development. Believing that they could successfully respect the wildlife protection policy while still assuring a means of living, a group of local people took the initiative of forming the Caohai Farmers Association to develop activities that would provide a boost to the district. The Association launched projects related to the management of the nature reserve, eco-tourism, publicizing Caohai culture, and providing technical training to promote livestock raising.

Although "wetlands conservation" is different from "forest restoration," there is a common purpose – to promote ecological conservation without a reliance on migration. This is why we felt that the methods developed by the Caohai Farmers Association could also be utilized in the context of *Tuigenghuanlin*.

Forest conservation by building community roads

The basic philosophy of the Caohai Farmers Association is that the success of ecological conservation depends, above all, on improving the quality of life of farmers and rebuilding their livelihoods. Although it may seem paradoxical, an afforestation project is more successful when money is spent on providing living infrastructure and microcredit to help farmers switch to alternative means of livelihood than when invested solely in afforestation. The first project that the Caohai Farmers Association initiated, at the request of villagers, was the construction of community roads. It is appropriate at this stage to ask how the construction of community roads leads to forest conservation.

When Mr. Deng Yi, a former secretary of the Caohai Farmers Association, and another young volunteer staff arrived in Gusheng Village, they first called villagers and arranged meetings at "block" *(zu)* level. A block is a naturally formed settlement. There are 15 blocks in Gusheng Village. Believing that an NGO would only provide funding for afforestation purposes, the villagers were surprised to learn that they could also access money for social infrastructure projects that they required themselves, such as roads, electricity and water supply. The residents of Blocks H1, H5, and H6 came together voluntarily and spent a night in discussion; as a result, they came up with a proposal to build community roads.

Many blocks in Gusheng Village have no roads through which cars can pass. For this reason, the cost of transporting coal is high, and large amounts of firewood are therefore collected from the collective forest. This in turn makes it difficult for the forest to expand. The absence of roads also makes it difficult to transport agricultural products out of the village. Even now, most of the farm households live more or less self-sufficiently. They have little desire to run a livestock-raising operation, and there is almost no investment activity. While it may appear that road construction has nothing to do with afforestation, improving the living standard of farmers by such measures makes them better disposed towards afforestation and forest conservation.

The NGO staff collected all the proposals of each blocks, wrote them down on a large piece of paper, and announced them publicly. The NGO emphasized the following four criteria for project adoption: (1) The villagers themselves are able to maintain and manage the project; (2) The cost calculations are rational; (3) Part of the cost is borne by the villagers themselves; and (4) The villagers clearly explain the relevance of the project to environmental issues.

At first, Blocks H1, H5 and H6, which strongly desired road construction, requested a large sum of money as compensation for use of their farmland and for labor costs involved in the construction. The NGO staff Mr. Deng Yi replied, explaining that

> The ultimate goal of this project is, as far as possible, and on the initiative of villagers themselves, to achieve both a better standard of living and environmental conservation. The NGO can only help a little by providing external financing. Since the funds available to us for the project are limited we can only cover a part of the construction costs.

The NGO asked each block to revise their funding requests. After some internal negotiations by each block, H5 and H6 decided that the farmers who would lose land to the road project would be compensated by an equal area of land contributed from each of the unaffected farmers. In addition, the farmers agreed that rather than seeking money for labor costs, they would supply labor for the project free of charge.

In addition to road construction, the farmers from H5 decided to seal off 100 *mu* of their collective forest to allow it to regenerate by means of *Fengshanyulin* (closing a mountain for restoration). The process of *Fengshanyulin* involves the prohibition of firewood collection and livestock grazing in badly degraded forest areas to enable regeneration of the collective forest.

In the end, H5 was granted the largest budget by the NGO, and the road-building plan proposed by H6 was also approved. H1 was unable to understand the purpose of the project, and the estimates that they presented were much higher than those of the other two Blocks. As a result, H1 was not granted any funds.

Feeling disappointed, Block H1 reduced the cost of their proposal, in line with the NGO's objectives of forest conservation, and also decided to set aside 1,500 *mu* of their collective forest land for *Fengshanyulin*. With its revised and more rational budgeting, H1's road construction proposal was approved one month following submission.

As shown in Table 4-4, after the road building projects of the first three blocks were accepted, further road building and the *Fengshanyulin* applications were submitted by other blocks. Even blocks that had not yet begun any funded projects began, on their own initiative, *Fengshanyulin* in order to help them obtain funds for future projects.

In the first term of the project, the ratio of funds spent by the NGO to that spent by the farming households was approximately 5:1, which suggests that the farmers were trying to appropriate as much money as possible from the NGO; however, in the second-term projects in Blocks H2 and H7, the ratio of NGO spending to farmer spending had fallen to approximately 1:1 and 3:2, respectively. This change reflects a change in consciousness of the villagers, who became more willing to spend their own money to achieve as much as possible for themselves.

Table 4-4. Road building projects and the *Fengshanyulin* in Gusheng Village

Project	Block	Period	NGO funds (yuan)	Farmers' self funds (yuan)	Governments' contribution	Fengshanyulin area *(mu)*
Road construction	H5	Nov, 2003	10,660	2,200	Explosives	100
Road construction	H6	Nov, 2003	8,600	1,700	Explosives	–
Road construction	H1	Dec, 2003	2,000	790	Explosives	1,500
Road construction	H2	Oct, 2004	3,900	4,200	Explosives	2,000
Road construction	H7	Nov, 2004	3,000	2,000	Explosives	100

After a period of one year, the *Fengshanyulin* initiatives had clearly had positive results. According to the reports of villagers, the ban on collecting firewood meant that the ground was covered with fallen leaves and dead branches, and mushrooms were seen for the first time in 20 years. The water retention capacity of the mountain increased, thus reducing the volume of muddy water flowing off the mountain on rainy days. Farming households were beginning to see hope in these collective forests, considering that in future they would be able to collect and sell mushrooms, using the income for ecological conservation.

Creation of a farmer's cooperative

Following the success of the road-building projects, the villagers established a management committee to oversee the projects, with members of the committee elected by popular vote. Compared with the current village leadership, it was relatively young people who were elected to the committee. The management committee members, together with the NGO, organize meetings of the villagers and select and manage the projects. The committee also manages the funds for the projects. The level of trust between the villagers and the NGO is growing steadily, as is the degree of participation of farming households in various projects. One farmer described the process as follows.

> At first, I didn't know the difference between an NGO and the government. In the sense that they bring money from outside to try and carry out a project in the village, they are more or less the same. I thought that it might be just a one-off thing, so if there was something to get, I would try to get as much as possible out of it. However, I found the NGO to be very flexible, and I realized that they were sincere in trying to help us. The NGO people stay in the village like other farmers. We can't deceive them. So, we, the farmers, do the best we can to help ourselves, but for whatever we really can't manage we call on the help of the NGO. If there is money left over, we can use it for other projects to improve the village.

Increasing numbers of people started to participate in the village meetings organized by the NGO, even if they had to take a day off work to do so. The meetings were very appealing to the villagers because they could discuss the projects and make decisions there and then. This was the villagers' first experience of independently planning and managing a joint project, and they felt a growing enthusiasm. In 2005, the management committee to oversee the projects became a cooperative, namely "Gusheng Village Farmers' Cooperative for Ecological Development."

Working with the government

For these projects to be successful, the villagers must not only work independently, but also engage the participation of the local township *(xiang)* or town *(zhen)* governments. To boost the district, a good working relationship between the villagers and local administrative bodies is an indispensable element. Thus, the NGO sought to rebuild a relationship of trust between the village and the local government.

The NGO presented the cost estimates and implementation plan of the road-building projects to the town government and asked for explosives to use in the project. In fact, the town government had wanted to build community roads within the village, but the project had never commenced because of the high cost of the work. When the town government officials saw the cost estimates presented by the NGO, they were extremely surprised. They were shocked to see that the total cost was only 10 percent of their own estimate for the government project. Accordingly, they happily agreed to provide the required explosives free of charge. The experience of this road-building exercise resulted in an enhanced level of trust and cooperation between villagers, NGOs and the government.

The government came to trust the technological capability of farmers' through the project. The government no longer tries to stop intercropping in new timber plantations, even though it is prohibited by the *Tuigenghuanlin* Ordinance. The government has observed that farmers can properly manage the plantation in conjunction with agroforestry practices.

Promotion of livestock raising

In response to a strong desire by Block D1 to expand their livestock operations, the farmers were provided with technical training in livestock raising and microcredit for the purchase of livestock. Funding was jointly managed by the project management committee and the NGO, and the project was implemented on a trial basis between October 2003 and April 2004.

At the time when the microcredit loans were due to be repaid, some farming households were unwilling to make the repayment. This also had happened with poverty support loans provided by the government. If the NGO had been operating alone, collecting the repayments would have proven difficult; however, the committee that manages the project finances is run by villagers, and any failure by Block D1 to repay its loans would prevent other blocks from receiving further microcredit. Consequently, the management committee visited Block D1, accompanied by representatives of Block H1 who were applying for microcredit loans in the following year. The committee members enquired as to why the loans had not been repaid, and asked the Block as a whole to determine a solution to the problem. As a result of this meeting, all outstanding loan repayments were made within a week.

The town government was so impressed with the NGO's method of operating that it sought to emulate its methods. Although the town government had provided poverty support loans a number of times, the poor rates of loan repayment had made it reluctant to provide further loans; however, the NGO's 100 percent loan repayment record gave the town government renewed hope for such a scheme. Consequently, after conducting a bottom-up survey of funding requirements and holding meetings with villagers, the town government issued a total of 28,000 yuan in poverty support loans to 28 households from Block H1.

In addition to purchasing livestock with the loan money, the villagers introduced a methane gas system. By fermenting livestock excrement and urine, the villagers were able to produce their own methane gas for use as fuel. This system provided an alternative energy source to the firewood that they had used previously, as well as providing organic manure.

Introducing non-timber forest products

One of the reasons that farmers were not keen on plantation management during the initial stage of *Tuigenghuanlin* was that they were not supplied with the kinds of plant varieties that they desired.

After many of the Yunnan pine seedlings in *Tuigenghuanlin* withered, the project management committee and representatives of the villagers decided to select other species to plant in the area. If villagers are able to independently select the commercial tree varieties to be planted, their enthusiasm for managing plantations could increase markedly.

One positive effect of establishing the project management committee was that villagers began to actively travel beyond the village to conduct surveys. In January and February 2004, a group of representatives of Gusheng Village made an inspection tour to Yunnan Province. After conducting their own market survey, they selected *Kuding* tea as a possible species to plant in the *Tuigenghuanlin* sites. To begin with, they bought 1,000 seedlings for trial cultivation in the village. After finding that *Kuding* tea was well suited to the climatic conditions of the village, they applied to the NGO for a loan to purchase more seedlings. As tea seedlings are expensive, the NGO and the farmers each contributed 50 percent to the cost of purchasing additional tea seedlings. Planting of tea shrubs in the afforestation area of the village began in November 2004.

Conclusions

We have described the anxiety felt by farmers following the introduction of the *Tuigenghuanlin* program, and reported on how NGOs launched initiatives that aimed to simultaneously regenerate forest and improve the lives of farmers, without relying on ecological migration. Although we followed the progress of these projects for little more than one year, we saw that over this time the villagers, who were previously passive, became progressively more active and positive. The following three points can be made regarding the significance of these projects.

Firstly, our observations confirm the effectiveness of the approach of the Caohai Farmers Association, who place an emphasis on respect for the opinions of villagers and begin by providing the most urgently required social infrastructure and then connecting this development to ecological conservation. There are many international aid organizations and NGOs that support afforestation efforts in China, but to be successful some of them need to change their approach. Instead of concentrating their funding only on afforestation, they need to implement programs that improve the lives of the affected people and enable them to take their own initiatives. Afforestation should be positioned as an extension of these programs. Just as the forest is a form of social common capital, so too are infrastructure and microcredit (Uzawa 2005). The social functions of the community are strengthened thorough the experience of collective management of such capital. The experience of building a community road or managing a microcredit program can thus contribute to building effective methods of forest management.

Secondly, before the implementation of these projects, each household was struggling in isolation to find a way to adapt to the changes brought about by the forest restoration policy. Most households appeared to have run into a dead end, finding themselves with no choice but to find work outside the village. With the commencement of the projects, the villagers were provided with a collective adaptation strategy to complement their individual household strategies.

It is said that there is strong tendency to individualism in Han Chinese agricultural village society, however, "individual" and "group" are not mutually exclusive concepts. Whatever the degree of individualism, the individual always needs a certain amount of support from social groups.

Following the commencement of the NGO activities, villagers began to hold frequent meetings and to collectively plan and manage projects for their blocks, including the construction of infrastructure and promoting livestock raising and tea shrub planting. Through these projects, villagers found renewed hope for their lives after the *Tuigenghuanlin* program. Previously, the collective forest of the village was "collective" in name only; in reality, individuals collected firewood as they pleased. With the advent of the NGO projects, each block began to manage their forest in a truly "collective" way, for example by establishing rules for utilizing the forest.

Thirdly, mediation by the NGOs transformed the relationship between farmers and the town government from one of growing mutual suspicion to one of mutual trust. The ability of the NGOs to objectively understand the positions of both the government and the farmers enabled them to skillfully mediate between the two parties. The three parties – the farmers, NGOs and government – were able to develop a stable, collaborative working relationship.

In China today, agricultural and environmental NGOs are relatively inactive, however, the initiation of spontaneous projects by grass roots NGOs at the village level can serve to complement government projects and help to ensure their success.

Additional Note

After the Japanese version of this book was published in 2005, the Chinese government decided on significant reforms of *Tuigenghuanlin* in August 2007. First, the government extended the period of compensation and subsidies for another eight years, to bring the total period to sixteen years. Second, the intercropping of soybeans and other crops with low stalks in forested land was legalised. The reforms are positive and have contributed to reducing some of the social unrest in upland communities and the possibility of re-reclamation of reforested lands. For farmers, this change in position is a victory over the government's inflexible policy.

On the latest development of the national *Tuigenghuanlin* program and also the settlement of NGO program conducted in Gusheng Village, we have published a book in 2009. See Yoshiki Seki, Xiang Hu and Narumi Yoshikawa, 2009, *China's Reforestation: Beyond Socialism and Market fundamentalism*, Ochanomizu Shobo (in Japanese).

References

Hirano Yuichiro, 2002, Gendai Chugoku ni okeru ryokukakatudo no tenkai to jumin sanka no seikaku ni kansuru kosatsu, *Hokkaido Daigaku Ensyurin Houkoku*, 59, pp. 67–98. {Japanese}
(Hirano Yuichiro, 2002, An analysis of the characteristic of people's participation toward afforestation programme in contemporary China (In Japanese, summary in English), *Research Bulletin of the Hokkaido University Forests*, 59, pp. 67–98.)

Tsurumi Kazuko, 1996, *Naihatsutekihattenron no Tenkai*, Chikumashobo. {Japanese}
(Tsurumi Kazuko, 1996, *Movement of the Theory of the Endogenous Development*, Chikumashobo.)

Uzawa Hirofumi, 2005, *Economic Analysis of Social Common Capital*. Cambridge University Press.

Xiang Hu and Seki Yoshiki, 2003, Chugoku no taikoukanlin to hinkonchiiki jyumin, Yorimitsu Ryozo hen, *Hakai kara Saisei he –Ajia no Mori kara*, pp. 149–209. Nihon Keizai Hyouron Sha. {Japanese}
(Xiang Hu and Yoshiki Seki, 2003, Land conversion programme in China and the fate of people in poverty, *From Destruction to Reproduction: Cases of Asian forests*, ed. Ryozo Yorimitsu, 149–209. Nihon Keizai Hyouron Sha.)

Zhongguo Kechixu Fazhan Linye Zhanlüe Yanjiu Xiangmuzu, 2002, *Zhongguo Kechixu Fazhan Linye Zhanlüe Yanjiu Zonglun*, Zhongguo Linye Chubanshe. {Chinese}
(Research Project Team on the Strategic Study of Sustainable Forestry in China. 2002. *General Theory Concerning the Strategy of Sustainable Forestry in China*, China Forestry Publishers.)

II Questioning economic aspects: Can "ecological migration" achieve a reduction in poverty?

The mechanism of poverty resulting from "ecological migration" 5

From case studies of herders in Minghua District, Sunan Yogor Autonomous County, Gansu Province

MAILISHA

Introduction

An area termed the "Heihe River Basin Ecological Preservation District" has been added to Chinese Government maps of Gansu Province in recent years. This district lies within the Gobi Desert, in the middle of the Hexi Corridor. Administratively, it is located within Minghua District in Sunan Yogor (Yugu) Autonomous County.[1] Several different ethnicities live in the area, including Yogors, Han Chinese, Tibetans, and Muslims, with Yogors making up 89 percent of the population. In recent years, with the implementation of environmental conservation projects,[2] a "national agri-

---

1   Sunan Yogor Autonomous County is home to multiple ethnic groups, the principal group being the Yogors. The county is located in Gansu Province within the central region of the Hexi Corridor in the northern foothills of the Qilian Mountains. The county can be divided into three zones, separated by other cities and districts. The population of the county is more than 35,000, with Yogors making up 28 percent and Tibetans 24 percent of the people. In addition to these two ethnic minorities, there are also smaller numbers of Mongolians, Muslims, Bao-ans, Dongxiangs, and Tus. In all, ethnic minorities make up a total of 56 percent of the entire population and outnumber Han Chinese in the county ("Sunan Yogor Autonomous County Statistics Yearbook," Statistics Bureau of Sunan Yogor Autonomous County, Gansu Province, 2001).
2   As a measure for conserving the ecological environment of the Heihe River Basin, the Chinese Government is currently modifying the characteristics of the population in the area. It is enforcing grazing bans in areas of cultivation forest that are water sources, and revising land utilization around the middle reaches of the river via the introduction of water-saving irrigation and ecologically efficient crop farming. In the Ejene (Ejina) River Valley, within the lower reaches of the Heihe River, the government is taking measures to relocate nomadic people and their livestock, thereby terminating nomadic livelihoods. This policy of adjusting the population distribution in the river regions has already begun.

cultural development district" has been established in these grazing lands and a policy of encouraging herders to change livelihoods has been implemented. However, many ecological migrants have been produced as a result of these policies. Although ecological migration measures have ostensibly been implemented as measures for providing poverty relief to ethnic minorities and for improving the lives of pastoralists, implementation of these measures has also been motivated by the need to facilitate the abovementioned projects. The incentives for inducing migration, particularly the migration of young families with children, include relief from poverty, the provision of wells, electricity, and water supply, and the construction of schools; however, while the relocated herders expect their lives to improve in the new migrant villages, it has transpired that approximately half of the heads of migrant households have to find work outside the village (Mailisha, 2004a). In addition, some of the ecological migrant families cannot send their children to school for economic reasons.[3] Another more serious problem is that violent conflicts have arisen over access to water supplies in migrant villages (Mailisha, 2004b). Consequently, numerous problems are emerging from the migrant villages that were previously intended to serve as dynamos for environmental conservation.

The villages have become the site of interconnected problems such as "resource plunder in the environmental conservation era" and "poverty caused by poverty relief policy."

Implementation of the migration policy has brought about great changes in people's lifestyles. Herders have been encouraged by the government to abandon their traditional forms of livelihood and to adopt the competitive attitude of free market economics. As a result, the herders have started to clear land for cultivation and have established livestock fodder bases ("man-made grasslands") to increase livestock production. As part of this process, resources that the herders had previously had little need of, such as natural resources, energy, and chemical fertilizers, have now become deeply incorporated into their daily lives. In seeking to establish a new livelihood by adopting a new lifestyle based on high levels of

---

3 When I conducted field surveys in 2002 and 2003, I encountered two instances of children leaving school before graduation. One was a case of two sisters and the other of a boy. The boy resumed his schooling after one year when local government officials called on his family and ordered him to return to school. The sisters, however, did not return to school and I was told that officials did not visit the home of the girls to enquire when they would be reenrolling.

resource consumption, the herders have been forced to spend beyond their capacity to generate income. The more that they are drawn into the competitive sphere of the free market, which values high-volume production and supply, the more they are under pressure to clear additional land and invest in additional facilities. Thus, new kinds of environmental problems are emerging in the very places where environmental conservation projects are being promoted. It is precisely here where the roots of the poverty problem lie hidden.

To maintain their new lifestyles, residents of the migrant villages have been forced to engage in economic activities in addition to their principle means of livelihood, such as clearing more land for cultivation and day labor. While this has kept unemployment in the area down, the financial situation of the inhabitants appears to be becoming increasingly difficult.

In this paper, I wish to elucidate the mechanism that has lead to these poverty problems.

Background to the implementation of the ecological migration policy

Minghua District in Sunan Yogor Autonomous County is located at the southern edge of the Badiinjiran Desert in the central region of the Hexi Corridor. The district borders the major agricultural region of Gaotai County in the northeast and Jiuquan City to the southwest. The district is a remote enclave, far removed from the capital of the autonomous county. The average altitude is 1,281 meters above sea level, annual precipitation is 66 mm, and the average annual temperature is 7 to 8°C. Within the administrative district are two inland lakes. Administratively, the district is divided into three townships and 14 villages. The inhabitants of Lianhua and Minghai Townships use the West Yogor language.

Up until the first half of the 20th Century, the people of Minghua District lived traditional lives as nomadic herders. However, in response to the steady expansion of "oases" in the surrounding areas, created by the clearing of land for cultivation, the herders began to live in settlements during the 1940s and the 1950s. The herders then engaged in the practice of grazing centered on their settlements and rotating their herds in the grazing lands of surrounding areas throughout the year. From early spring until summer they continued the practice nomadic herding and traveled to remote parts of surrounding regions and to desert zones. However, over several decades, vast expanses of grazing land in Yogor Autonomous County have been lost because of the large-scale clearance of land in surrounding agricultural regions. In addition, with the introduction of the "contract production" system in 1982, the ownership of land and livestock was privatized. Since this time, the grazing lifestyle of the people, which relies heavily on the natural environment, has become increasingly difficult.

A significant underlying component of the ecological migration policy in Minghua District is the problem of environmental degradation of grasslands.

Minghua District has been historically been blessed with abundant rainfall. The two townships of Lianhua and Minghai take their names from the two lakes in the area. In addition, groundwater conditions were also good. According to the accounts of elderly local herders, up until the 1950s groundwater could be found by digging to a depth of just one meter and wild animals were also numerous at the time.

Partly because of droughts that lasted for several years, the "lakes" dried up and groundwater levels fell. As evidence of this trend, many families now use their dry wells as "refrigerators." Given the deterioration of the grasslands, the area of usable grazing land in Minghua District is now approximately equal to what it was in the 1950s.

It is said that the primary cause of this kind of environmental degradation is anthropogenic, rather than natural. In the agricultural zone along the upper reaches of the Minghua District, large-scale irrigated agricultural development has been pursued relentlessly for many years. Since the establishment of the People's Republic of China, the government has repeatedly relocated people for the sake of national resource development. One of the principal destinations for these migrants in northwest China has been the oases in the Hexi Corridor, which have now developed into a national-level "food base" and seventy percent of food production in Gansu Province originates here. According to the herders of this area, excessive extraction of groundwater in these agricultural areas is the main cause of falling groundwater levels in Minghua District.

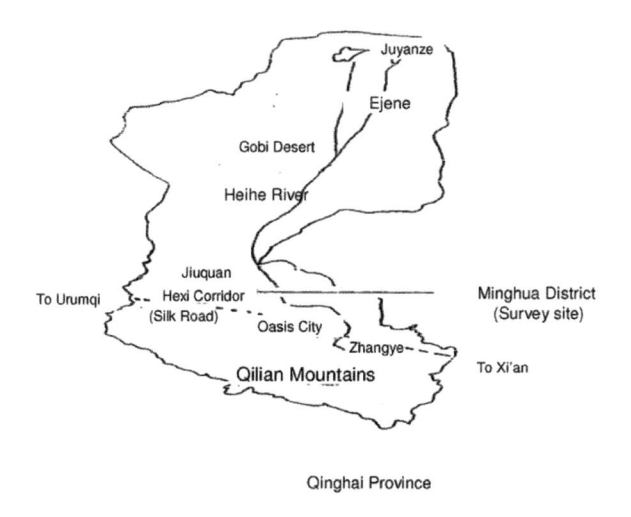

The mechanism of poverty resulting from "ecological migration"

Fig. 5-1. Location of the survey site in the Heihe River Basin

Fig. 5-2. Minghua District, Sunan Yogor Autonomous County, and surrounding areas

The environmental conservation of grasslands is an urgent problem that has ramifications for environmental protection across China as a whole and must be addressed. While the herders have long sought action to stem the destruction of the grasslands, they could never have imagined the environmental conservation measures eventually implemented by the government. While the herders had high expectations of being able to regain their lost grasslands, the result was that they have had to give up even those few grassland areas that they had somehow managed to maintain.

In the latter half of the 1990s, a "national agricultural development district" was created in the grasslands of Minghai Township, Minghua District. The herders in the area were forced to live together in concentrated settlements and contribute to agricultural development work. To date, large numbers of Yogors, Tibetans, and Han Chinese were relocated to the district from the middle and upper reaches of the Heihe River within the county and there are already three migrant villages here. One of these, Shuanghaizi Village, is inhabited by people relocated from Lianhua Township. It was in this village that I conducted a survey on the efforts of these people to change their means of livelihood.

Since 2000, Sunan Yogor Autonomous County has been implementing a policy of relocating some of the herders from the Lianhua Grasslands to the "national agricultural development district" to engage in agricultural development. The official aim of this policy was to reduce the burden on the grasslands and also to alleviate poverty among the herders. Through this policy, the government is seeking to reduce the population of the Lianhua Grasslands to enable the vegetation to regenerate.

Photo 5-1. Newly built migrant village (agricultural development district)

According to a document related to the county's "Ecological Environment Conservation Plan" (Sunan-Xian Renmin Zhengzhi, Xianwei, 2000), the herders' decentralized style of living and traditional extensive grazing methods place excessive pressure on the ecological environment. This has not only caused destruction of the ecological environment, but also significantly limited the economic development of the livestock industry in the area. The document states that, accordingly, the herders are to be removed from the degraded environment and relocated to a site with better environmental conditions and transport infrastructure. In it addition, purports that, in addition to being an effective mechanism for rapidly relieving the poverty among herders and helping them to live a richer life, the plan will also be beneficial for the ecology of the environment.

To date, approximately 70 herder households, approximately half the Lianhua Township population, have been forced to migrate to the Shuanghaizi migrant village in the agricultural development district. Most of the migrants are young and middle-aged people aged in their 20s to 40s. Conversely, those who have not migrated are primarily the elderly, families with sick or mentally and physically challenged members, and families without male children. It is possible that households with young males were more positive about migrating because they wished to secure a house in preparation for marriage.

Efforts of ecological migrants to change their livelihoods

Changing from livestock raising to crop cultivation: Case Study 1

Family "A" (resident within an ecological migrant village in the "national agricultural development district") consists of six members: Mr. A (aged 44), Mrs. A (37), their two daughters, and twin sons. In 2000, the family sold 150 goats and left the Lianhua Grasslands to live in the migrant village. Given their haste to relocate, they had to sell their goats at the low price of 80 yuan per head. They say that they moved because the area in which they lived previously was too remote, because they wanted to take advantage of the preferential government treatment given to the new settlement, and to take advantage of schooling for their children. When the family lived in Lianhua, the school was so far away that Mrs. A lived with the children near the school, forcing the family to live apart.

To address the sense of dissatisfaction among migrants when the migrant village was built, the government promised that their standard of living would improve after relocation. To meet this promise, the government invested in basic services and facilities such as water supply, electricity, wells, and roads. In addition, the government has paid each migrant household from Lianhua 10,000 yuan to subsidize the construction of a house. This subsidy is not paid to households in the other two migrant villages within the agricultural development district.

Using the money generated by selling their livestock when they left the grasslands, the subsidy from the government, and borrowed money, families are expected to build themselves a house and clear land for cultivation. The migrants have already moved into their houses and begun to cultivate crops. Family A has cleared an area of 30 *mu* (1 *mu* = 0.67 hectares), and are now growing wheat, corn, wheat for brewing beer, and grass.

Income and expenses of relocated herders

From one *mu* of land, Family A earns approximately 200 yuan per year, however, it costs them an average of 170 yuan per *mu* to raise their crops. In addition, they are not self-sufficient and have to purchase food for consumption. Unlike the grazing of livestock, crop farming is very costly. In many cases, the investment in electricity, fertilizer, agricultural chemicals, and equipment makes crop farming unprofitable. For this reason, Mr. and Mrs. A have had to take on various additional jobs, such as becoming day laborers. The government is developing the agricultural development district not only for the resettlement of migrants, but also to attract private industry from outside the region. For example, companies are currently developing a 10,000-*mu* grape plantation and a 10,000-*mu* medicinal herb farm.

I learned from representatives of the companies that despite the large areas of cultivated land they possess in the region, only management personnel are sent to work here. The actual farming work is undertaken by labor employed on-site. This shows that ecological migrants have inadvertently become a source of cheap labor for privately owned external companies. The family's financial situation meant that the two daughters (aged 16 and 14, both junior high school students) had to quit school in 2003, and 2004. While the daughters had wanted to continue their studies, the money spent on their schooling could be used to buy substantial amounts of agricultural chemicals and fertilizer. Unless the family buys enough agricultural chemicals and fertilizer, they cannot produce enough from their land and the livelihood of the entire family is then threatened.

Thus, while family A came to the migrant village for the sake of their children's schooling, ironically, as a result of their relocation, their children can no longer attend school. In the grasslands where they had lived previously, children were never prevented from attending school due to poverty. Now, the two daughters are working away from home and send the money that they earn back to the family.

Most of the migrants from Lianhua Township are now burdened with debt. According to statistics collected by the County Agricultural Office (Sunan-Xian Nongye Bangongshi, 2003), the average debt per household in Minghai Township is 6,650 yuan. Under pressure to pay off their debts, more than half of the heads of migrant households have either taken on outside jobs, or temporarily pursue livestock grazing in the grasslands where they had previously lived.

In the future, the A family hopes to save money and grow grass on their arable land and raise livestock. In fact, in this area, an "enclosed livestock raising" project has been launched, in which an area of grassland is fenced off to raise livestock. Some households in the migrant village have taken bank loans to raise cattle, and an increasing number of people are taking outside jobs to raise money for this purpose. However, the leadership of the migrant village is already worried about the market principles that govern the distribution of livestock products. There are currently wide price fluctuations in the Chinese market for livestock products; in times of excess supply, such products can be hard to sell, resulting in low returns. It is likely that for migrants to engage in "enclosed livestock raising", they would need to take on further debt, however, if they were to face unfavorable market conditions, they would experience more severe financial difficulties, possibly beyond the point of recovery.

The biggest change for the migrants is that each house in the village has its own water supply. In terms of the ready availability of water, life is much more convenient that it was in Lianhua.

For farmland irrigation, there are deep mechanical wells (150 m deep) in the village for shared use. One of the wells is shared by 16 households. In summer, water is pumped from the well 24 hours a day. Deep wells like these cost approximately 130,000 yuan, but construction costs were covered by the government as part of the facilities provided to the migrants. Construction of a total of 150 such wells for irrigation is planned for the agricultural development district.

From time to time, power blackouts interrupt the supply of water, leading to great confusion. Electricity used to power the wells is also expensive and must be paid monthly; many households fall behind on the payments. After a notice urging settlement of the electricity bill has been posted in the village, households failing to pay have their water supply cut off from the following month. Beginning in 2004, the system of payment was set to become even stricter, such that if even one household fails to pay its share of the cost, supply would be cut off to all 16 households using the well.

Although migrant villages were created for the purpose of environmental conservation, they are actually having an adverse effect on the natural environment, as they have resulted in increased consumption of water, energy, and natural resources. In addition, as Minghai Township is located upstream of Lianhua, agricultural development in the "new settlement" in Minghai is causing groundwater levels to decrease in the area from which the migrants relocated. These observations demonstrate that the reality of migration is that instead of stemming environmental degradation, increased consumption of resources has resulted in increasing the extent and rate of the destruction.

Shifting from extensive grazing to intensive livestock raising: Case Study 2

I relate the case of the family B, who have decided to remain in an area from which people migrated. Shangjing Village in Minghai Township is the largest village by area in Minghua District. Long ago, the herders of Shangjing Village had the largest area of grazing lands in the region. With the establishment of the national agricultural development district, the herders lost a vast amount of their common grazing lands. As a result, they could not sustain a livelihood by grazing in natural grasslands, and were forced to change their method of livestock raising.

Family B has four members: Mr. B, (aged 50), his wife (54), and two sons (24 and 16). The eldest son works in the municipal offices of the township as the accountant of the village and lives in the center of the township. The younger son is a high-school student who lives in a dormitory at his school. A niece of Mr. B's (15) came to live with the couple to help them with their daily tasks. She receives 200 yuan per month from Mr. and Mrs. B.

At present Mr. B keeps 150 goats, 89 sheep, 25 cattle, 3 pigs and 30 poultry. He has also expanded his farming land from 5 *mu* at the time that the "contract production system" was introduced to 52 *mu*. On his arable land he grows agricultural products and grass for livestock fodder. Mr. B has 35 *mu* of wheat under cultivation, 7 *mu* of corn, and 15 *mu* of feed grass. With a yield per *mu* for wheat of 350 kg, he produces a total of 10,500 kg of wheat per crop. Of this he sells 4,500 kg for a return of approximately 4,500 yuan. The corn is used for livestock feed, and the wheat and corn straw are used as coarse feed. Mr. B manages to reap two crops of feed grass per year and produces more than 5,000 kg of grass per year from 15 *mu*. In this way, using their farmland as a base for production, family B has attempted to expand its livestock operations. Mr. B can depend on natural grasslands to feed his livestock for only five months of the year. For two months, May and June, he grazes his animals in grassland near his home, and for three months, from July to September, he travels widely with his herd and lives in a tent, grazing in common grasslands. For the remainder of the year, he feeds his livestock with hay cut from natural grasslands, cultivated feed grass, straw from his crops, and corn. Most of the herders of Shangjing Village follow much the same pattern of farming as that of Mr. B.

Photo 5-2. Imported breed-improved sheep are popular for their high reproduction rate

Some other households are also increasingly cultivating and grazing their animals in remote areas. One example is the ex-deputy mayor of Minghua District, who cultivates a total of 60 *mu* in three different remote areas and maintains a house and livestock sheds in each area. Since 2004, he has been growing feed grass at these farms and rotating his herd between them according to the season.

Most of the herders of Shangjing Village expanded their area of cultivated land during a land-clearance boom. Along with the development of agricultural land and an influx of migrants to the area, the inhabitants of Shangjing Village were encouraged to grow livestock feed crops. Seventeen households collectively cleared a large square-shaped area of land for cultivation, dividing it into household parcels of an average area of 23 *mu*. Mechanical wells were constructed for irrigation, with 71 percent of their cost subsidized by the national government.

All the herders of Shangjing Village now have large areas of land under cultivation. Using these large areas of cleared land to produce feed crops, they are seeking to run increasingly larger livestock operations. As they rely on irrigation to grow large quantities of feed crops, water resources in Shangjing Village are becoming increasingly exploited, as in the agricultural development district and the migrant villages. With the aid of irrigation, it has become possible to reap two crops per year: a summer crop and a winter crop. While irrigation increases livestock productivity, ironically it also works to undermine its sustainability. With the large amounts of energy invested in cultivated land, large amounts of water pumped from the ground, and the use of chemical fertilizers, the natural resources of the area are being increasingly abused and depleted. Already, Shang-er Lake in Minghai Township has dried up, and groundwater levels in the area are falling significantly.

Income and expenses

Mr. B's income from agricultural products for 2004 was 4,500 yuan and was earned entirely from the sale of wheat. Total income from sales of livestock was 9,000 yuan: 10 goats for a total of 2,100 yuan, 10 head of cattle for a total of 6,400 yuan, and 1 pig for 500 yuan. Total income from the sale of livestock products was 4,250 yuan: 3,200 yuan for cashmere, 150 yuan for sheepskins, and 90 yuan for sheep wool. Mr. B's total income for the year was 17,750 yuan, which is considered high and indicates that raising livestock on feed crops potentially offers significantly higher financial gains than herding on natural grasslands.

The herders of Shangjing Village have the highest income in Minghua District. As a result, it is expected that the production methods of Shangjing herders will be promoted among other herders in the district as a model of modern livestock-raising methods and ecological environment management. Since 2004, grazing has been banned in Minghua District. According to the document, "Strategic Plan for the West Development Project" (Sunan-Xian Renmin Zhengzhi, Xianwei, 2000), over the next five years livestock production methods are to shift gradually from natural grazing to "semi-enclosed livestock raising" or "fully enclosed livestock raising," and high-productivity grasslands and fodder bases are to be established. The strategy aims to reduce the burden on grasslands caused by extensive grazing, to harmonize economic, social, and ecological factors, and to strive for sustainable development.

However, the cost of maintaining a lifestyle like that of Mr. B is extremely high. While his annual income for the previous year was 17,750 yuan, he had to pay approximately 11,600 yuan to cover the costs of feed crop production, including electricity, fertilizer, agricultural chemicals, and fuel for machinery. In addition, he paid 1,100 yuan in land tax and livestock tax, and approximately 2,000 yuan in wages to his niece. After subtracting all these expenses, his net income was comparable to that of most herders who continue to live in the traditional way, in some cases lower. In other words, while the scale of his operations expanded, it did not lead to a higher net income. For the B family, it appears that the more that they move away from tradition extensive livestock grazing toward modern livestock management, the tighter their financial situation becomes. In fact, to pay for the schooling of his children, Mr. B has had to seek additional work outside grazing and feed crop cultivation. He uses his own truck to help others to cultivate their land. Last year Mr. B undertook a total of 200 days of such additional work in spare time between busy periods, earning a total of 5,000 yuan. In this way, Mr. and Mrs. B have expanded their sources of income by combining grazing, feed crop cultivation, and day labor. Caught in a vicious circle of excessive investment and high workload, Mr. and Mrs. B are physically struggling to keep pace with the demands of their livelihood.

An alternative model for environmental conservation

In August 2004 I revisited Lianhua Township in Minghua District, from where many ecological migrants originated. With implementation of the ecological migration policy, many young people moved away from this area to settle in a new migrant village, and the herders who have remained have continued their traditional way of life as pastoralists. While the economic productivity of these herders, who live and work in coexistence with nature, is low, they are able to limit their expenses to less than 10 percent of their income and are thus financially stable. This represents a clear economic virtue of the traditional way of life. In the past two or three years, demand for livestock raised in Lianhua has been strong. Without becoming caught up in free market competition like those living in the development district, herders here can sell directly to buyers who actively seek the livestock that they raise. Some herders have even cultivated their own distribution channels to get their livestock to market. Here I present one such example.

Mr. C (aged 49) is a herder with a family of five residing in Lianhua. Since he has only daughters, he chose not to relocate to a new migrant village. At present, he sells goats and sheep manure. One of his daughters who works at Minzu Park in Jiuquan told him that the park wanted livestock manure from pollution-free grasslands to use as fertilizer. When he was transporting sheep manure to Jiuquan for the first time, several people he met on the way asked him about the manure and wanted to buy some. Although he couldn't sell the manure he was carrying, the experience gave him a surge of confidence in his "product."

Mr. C has now entered into contracts to supply his livestock manure to Minzu Park and Beijiao Park, both in Jiuquan, and he now also collects manure from other herders. Given he pays more than other buyers, the herders are happy to supply him with their manure. Mr. C now has a network of friends in Jiuquan from whom he receives information, including contacts who are seeking supplies of meat and livestock. Some of these people pay more than double the normal price he receives from commercial buyers. On the subject of these higher prices, Mr. C says, "If someone would buy all our products at such a price, herders could even reduce their livestock numbers but still have a more secure livelihood."

The notable point here is that this kind of independent distribution network takes on a positive significance in terms of the environmental conservation of grasslands. By opening up suitable distribution channels, products created using natural production methods can be connected with consumers who are concerned with safety, thereby establishing an environmental conservation system of sorts.

China has entered into an era of market economics characterized by mass production and mass supply; even in livestock feeding there is increasingly severe competition. The production of goats and sheep using feed rather than natural grass is placing pressure on traditional livestock-raising methods. At the same time, consumers are increasingly questioning the health implications of methods that force-feed livestock to fatten them over a short period of time. The psychology of Chinese consumers, who once sought only inexpensive products, is now changing, and there is a significant trend to valuing quality ahead of quantity. Reflecting this change in consumer attitudes is the appearance in supermarkets and shops of value-added products such as "pollution-free, natural grassland meat." Amid the dominance of mass production, the value of livestock raised in natural environments is becoming increasingly recognized as a high-quality, precious commodity.

Despite these trends, the livestock products of herders are still bought at low prices by commercial buyers. This is not because the herders have been left behind by the market economy, but rather, because the market is unstable. In this sense, there are still latent possibilities in traditional subsistence patterns that coexist with nature. If environmental conservation policy is approached from a different angle, it might prove to be more effective. For example, the government could become involved in distribution. If a distribution system were created that is mutually beneficial to both herders and consumers, the traditional system of grazing could be maintained. In this way, the herders could live a more secure life, and environmental conservation would also be possible. However, in Lianhua Township, a total grazing ban is expected to be implemented in the near future, in which case the inhabitants will likely be forced to move away.

References

Mailisha, 2004a, Seibudaikaihatsu no naka no shosuminzoku seitaiimin- Shukunan Yuguzoku jichiken ni okeru chosahoukoku, Aichi Daigaku Gendai Chugokugakkai hen, *"Chugoku 21"*, 18, pp. 79–86. {Japanese}
(Mailisha, 2004a, Ethnic minority immigrants under the western region development: A report from the sunan Yogor Yutonomous County, Aichi University, Faculty of Modern Chinese Studies, *China 21*, (18), pp. 79–86.)

Mailisha, 2004b, Kokuga churyuiki ni okeru ningenkatsudo to mizuriyo, Sogo Chikyu Kankyogaku Kenkyujyo Oashisu Purojekuto Kenkyukai *"Oashisu Chiiki Kenkyu Kaiho"*, Vol. 4, No. 1, pp. 53–72. {Japanese}
(Mailisha, 2004b, Human action and utilization of water resources in the midstream of the Heihe-River basin, Research Institute for Humanity and Nature, *Project Report on an Oasis-region*, 4(1) pp. 53–72.)

Sunan-Xian Renmin Zhengzhi, Xianwei, 2000, *Sunan-xian Yugu-zu Zizhixian Xibudakaifa Zhanlüe Guihua*. {Chinese}
(County Committee of the Sunan County Government, 2000, *Strategic Plan for the West Development Project in Sunan Yogor Autonomous County*.)

Sunan-Xian Nongye Bangongshi, 2003, *Sunan-Xian Minghua-Qu Shengtai Huanjing XianZhuang yu Jingji Kechixu Fazhan Zhuangkuang Diaocha Baogao*. {Chinese}
(Sunan County Agricultural Office, 2003, *The Research Report on the Present State of Ecological Environment and Sustainable Economic Development in Minghua District, Gansu Province*.)

The effectiveness of "ecological migration" in reducing poverty (1)

A case study based on the Tarim River Basin, Xingjiang

LI JINGYI

Introduction

The Tarim River is the longest and largest inland river in China. With a total length of 2,349 km and a main course of 1,321 km, the river is widely regarded as one the world's great inland rivers. The total catchment area of the Tarim is 1.02 million km², of which 996,000 km² lie within China.

The Tarim serves as a life source for the people of southern Xingjiang and protects the ecological environments of oases (Fig. 6-1).

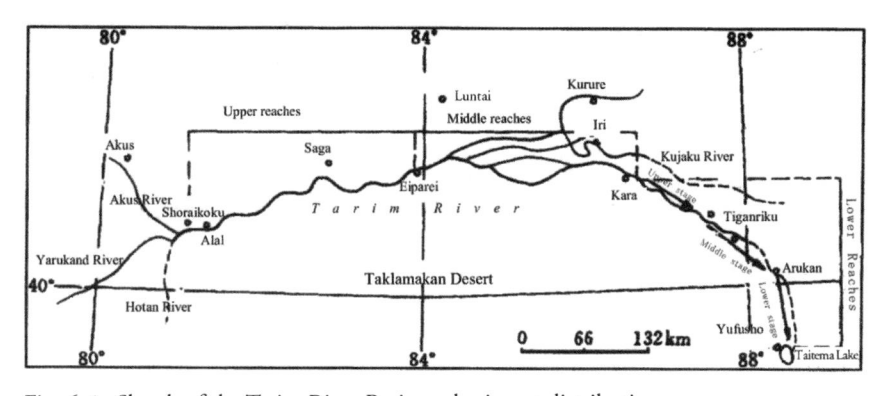

Fig. 6-1. Sketch of the Tarim River Basin and migrant distribution

Over the past 20 years, the effects of human activities and climate, along with poorly orchestrated development initiatives and inefficient utilization of water resources have resulted in a steady decline in the quantity and quality of water flowing in the Tarim River. Flow has ceased completely in the lowest 400 km of the river, causing Taitema Lake to dry up and large areas of poplar forest growing along the river's main course to wither. In the 1950s, the total area of poplars within the middle and upper reaches of the river was 400,000 ha, but by 2001 only 240,000 ha remained. Over the same period, the total area of desertified grassland increased from approximately 27,000 ha to in excess of 110,000 ha (Xingjiang almanac, 2001). As a result, the "oasis corridor" separating the Taklimakan and Gurbantünggüt is on the verge of disappearing; the two deserts having now almost joined together.

In respponse to the natural geographical conditions and the impact of human activities along the Tarim River, the ecological environment of the river's lower reaches is deteriorating daily. Water resources have almost been completely depleted and this has placed severe limitations on sustainable development of the social economy and human subsistence patterns. The threat of this problem to the ecosystem of northwestern China has attracted considerable public attention and resulted in government intervention in the form of environmental management initiatives. Ecological migration, in combination with poverty relief, has been implemented and is viewed as an essential component of government efforts in the areas along the middle and lower reaches of the Tarim.

In 2001, the Chinese Government invested seven billion yuan in environmental projects, including a comprehensive management system for the Tarim River, "cultivation discontinuation for enclosed forest restoration," and resettlement of inhabitants in order to remedy the aforementioned environmental problems. Ecological migration was implemented in Luntai and Weili Counties, Bayangol Mongolian Autonomous Prefecture, Xingjiang (a total area of 1.02 million km^2). This measure included pumping 1.5×10^8 m^3 of water from the Tarim River and transferring it downstream to enable water to reach the lower river reaches. This measure aims to reduce the destruction of natural poplar forests by people and livestock and to conserve water and improve water utilization efficiency to protect a total of 98,000 ha of natural poplar forests and shrub lands along a 100-km stretch of river. This area is intended to serve as a natural defense mechanism to protect the ecology and production of the natural poplar forests. As the degraded ecological environment of the affected areas are regenerated and become rehabilitated as much as possible, the living conditions and ideas of the herders who inhabit the area can be improved. In addition, by revising the industrial structure of the area, herders will have the opportunity to increase their income over a short period of time. Through the combination of ecological, economic, and social effects of this initiative, the full potential of this area can be realized and sustainable development of the area's economy will occur. As Uigur nationality constitute a large proportion of the ecological migrant population of this area, the question of the ethnic characteristics of minority ethnic groups is of special importance in research into ecological migration along the Tarim River.

Overall structure and implementation of ecological migration along the Tarim River

Necessity

(1) Ecological migration for poverty relief

When expressed as a percentage of the population, Xingjiang has more people living in poverty than any other area of northwestern China; this is due to the disparity of economic development in the area. It is striking that the majority of Xingjiang's economic poor are living in areas of ecologically poor environments. When it comes to policies on poverty relief, it is sometimes the case that twice the effort results in half the effectiveness. This implies that even with significant efforts, achieving poverty relief is often considerably difficult. There exists a real need to improve the undeveloped conditions of this area, and ecological migration has been proposed as the the only viable way to achieve this aim. In other words, it is considered essential to remove the poorest segment of society from unsuitable areas and relocate them in a more productive environment that has improved living conditions. It is thought that in their new environment, the migrants will enjoy a tranquil life, and by devoting themselves to work, will be able to escape poverty and improve their standard of living.

(2) Ecological migration for environmental protection

The environment is a complex system that sustains life and is comprised of a diversity of elements built up over a vast period of time. Similarly, environmental destruction can be considered the result of a large number of factors acting over long periods of time. Of these, one of the most important contributing factors leading to environmental destruction is population growth. In recent years, the impact of people living alongside the Tarim River has resulted in an annual decrease in water levels and the area of forest and grassland along the middle and lower reaches of the river has steadily decreased. Concomitantly, there has been an increase in the spatial extent of desertification is increasing in size, and the natural environment around the lower reaches of the river is rapidly deteriorating. The implementation of ecological migration will enable an improvement of the environmental condition of the river.

(3) Ecological migration for tourism development

Ecological migration is an extremely effective way to restore natural environments to the original beauty that they possessed as tourist attractions. After implementation of the migration policy is complete, regional and local governments plan to create "poplar forest parks in the desert" within "environmentally restored" districts along the middle and lower reaches of the Tarim River. These parks will attract tourists and generate revenue for local public bodies.

Favorable conditions and leadership ideas

Favorable conditions

A review of the state of migration along the Tarim River shows that most ecological migration projects have been conducted with the involvement of both government and the private sector. In 2001, the national government contributed a total of 110 million yuan to various Tarim River migration projects, and launched migration-related construction projects in 22 locations. While the government is using political means to carry out the ecological migration policy, it is also making use of free market forces by offering the incentive of profit-making to attract the participation of relatively well-established private companies. This approach provides relocated herders with an opportunity to increase their income by working for such companies.For example, the Ruiyuan DAIRY CO. LTD in the Bayangol Mongolian Autonomous Prefecture contributing knowledge to local people concerning the Raising and breeding of Holstein cows imported from the Netherlands, and providing calves to migrants free of charge. The objective is to help ensure that migrants can establish a normal livelihood and lifestyle following relocation. After training the migrants to Layisue the dairy cows, the herders are left to tend the highly productive Holstein cows themselves. Raw milk from the cows is supplied exclusively to the dairy group at a price that is favorable to the migrants. At well as assuring a reliable supply of high-quality milk to the dairy company, the initiative also helps the migrants to generate income and establish a new livelihood. Nonetheless, as the participation of private companies in this way is based on the opportunity for the company to make a monetary profit, it is necessary to provide government guarantees to protect against economic or market fluctuations.

Leadership ideas

The Tarim River Ecological Migration Projects are an opportunity to fulfill the aims espoused in the slogans, "Great western development" and "Construction of a great northwestern area with beautiful mountains and rivers." The projects must also adhere to the following six principles set down in the "Tarim River Basin Short-term Comprehensive Management Plan."

(1) Protection and improvement of the ecological environment is the primary objective. Ecological migration and environmental protection is to be closely coordinated with the systematic development and efficient utilization of natural resources, and a balance is to be achieved between ecological, economic, and social effects to realize harmony between people, nature, and sustainable development.
(2) Migrants are to be relocated principally to large-scale agricultural divisions. Based on the principle of development migration, a strong emphasis is placed on planning. Projects should proceed in stages, according to associated extent of difficulty beginning with the easiest stages.
(3) The comprehensive deployment of natural resources will be improved. Ecological migration and oasis development, desert environmental protection, and environmental improvements are to be combined.
(4) Living standards should improve following relocation. At the same time, resource allocation should be based on local social and economic conditions.

(5) Projects should be economically profitable and beneficial to the livelihood of migrants. Sufficient respect should be given to the traditional subsistence patterns, lifestyles, and popular customs of all ethnic groups.
(6) Current resources in neighboring regions should be surveyed and appropriately utilized. Ecological migration and industrial development should be linked to poverty relief for peasants and regional economic development to realize sustainable socioeconomic development.

Implementation process

Comprehensive plan

In the space of four years, 2,420 people from 524 farming households from a single township and two farms (agricultural and livestock) on the north shore of the Tarim River, Caohu District, Luntai County, migrated to Layisu District in the Dina River Valley (part of Qunbake Town, Luntai County, 21 km southwest of the county center) and Lapa District (part of Dadaonan Town, Luntai County, 11 km southeast of the county center). A total of 620 people from 124 households settled in Layisu District, while a total of 1,800 people from 400 households settled in Lapa District. Of these, 495 people from 42 households relocated between the end of 2001 and 2002 (first migration period), 429 people from 100 households relocated in 2003 (second migration period), 751 people from 179 households relocated in 2004 (third migration period), and 1,045 people from 203 households relocated in 2005 (fourth migration period). After the migrations had been completed, all 4,000 ha of the land in Caohu District, Caohu Agricultural Farm, and Karqiaqi Livestock Farm were subject to "cultivation discontinuation for forest restoration."

Migrant placement method

The choice of method for placing migrants is extremely important, as different placement methods have different associated social and economic effects. Selection should consider not only resources and socioeconomic development factors, but also individual factors such as age, education level, and psychological adaptability.

Thus, for ecological migrants from the Tarim River, the government adopted a placement scheme that was readily accepted by farmers and herders: a "household together"[1] scheme that encouraged herders to apply for migration voluntarily and organize themselves independently. This scheme promoted "development migration" to targeted areas. At the same time, the government proposed a development model that combines primary, secondary and tertiary industries. In addition to providing conditions suitable for production, the migration districts are prepared using infrastructure-equipped residential areas by utilizing the advantageous conditions of (existing) villages and towns.

Relocation characteristics

(1) Migrants feel a certain amount of unwillingness regarding migration

The majority of migrants are motivated to migrate by the desire to improve their livelihood and living conditions. As serious consideration is given to personal choice in the process of selecting a migration destination, willingness to migrate is a common characteristic of the Tarim River ecological migrants. However, as the entire community is moving from their original settlement, there is a certain degree of unwillingness among a small number of migrants.

(2) Relocation is within the same general area and occurs over several stages

Due to funding and geographical factors, as well as the limitations of ethnic customs, ecological migration within the Tarim River area is implemented within the county of origin. There is no relocation across districts, and migration is implemented in several stages.

(3) The dual objectives of poverty relief and ecological protection are combined

Ecological migration does not consider only the two layers of ecological protection and migration – it also incorporates the important element of poverty relief.

1   In one method of migrant relocation, a certain number of households are relocated together to establish a section within the new settlement.

Analysis of the effectiveness of ecological migration in the Tarim River Basin

Ecological effectiveness

Ecological migration has a great influence on the protection of the natural ecological environment and the protection of biodiversity. Although there are relatively few migrants in the Tarim River Basin, they have had a substantial impact on the area. Villages here were originally scattered around the area like forts, but they have now disrupted the habitats of wild animals and vegetation and currently pose a major threat to the existence of wild vegetation. Once these villages are demolished, the entire area will be transformed into a natural ecological system and a paradise of wild plants that will be free of human interference.

There have recently been marked improvements in the ecological environment. With the continued departure of inhabitants, and with large amounts of river water being transferred to downstream areas on five separate occasions, signs of life are appearing within the lower reaches of the Tarim River for the first time in over 30 years. The area of poplars is expanding, increasing from a low of 9,333 ha to 14,000 at present. The diversity of desert vegetation has increased markedly on both sides of the river. Reeds have reemerged in the marshes of Taitema Lake, and bighead fish from the Qarqan and River perch from Bosten Lake are beginning to reappear. On the river's shores, flocks of ducks, cormorants, and *wild goose* (a type of goose of the Tarim area that is smaller than a eagle but larger than a crane, weigh approximately 7–10 kg) are commonly seen watering in formation. All this activity is like the singing of a paean to the oasis.

Economic effectiveness

Ecological migration resulted a dramatic shift in the traditional subsistence pattern of the migrants, which principally involved livestock herding. Changes in two areas – the pattern of traditional subsistence and the industrial structure of the Uigur – have increased economic prosperity. Furthermore, relocating inhabitants into concentrated settlements has aided the development of private enterprise in towns and townships, resulting in increasing tax revenues in the region. These conditions have brought increased incomes and new development opportunities to the Uigur people via a reorganization of extant industrial structure. Many industries are now able to grow rapidly, and these are expected to become new high-growth areas in the economy of Luntai County.

The incomes of farmers have been increasing in recent years; consider one migrant, Ali Yusuf. In 2002, his family took the initiative of selling melons at the roadside of *Tazhong* Highway. The family's daily net income from this activity alone was over 170 yuan, generating an increase in annual income by 15,000 yuan. By 2003, inspired by Ali's success, a further 17 households of herders and farmers in the migrant village were engaged in selling melons. The migrant families not only sell their unique Caohu melons in large cities as far away as Guangzhou and Shanghai, they also help to promote the restructuring of local industry. According to a 2003 survey by the Luntai County Statistics Office, the total area of melon cultivation in Caohu Township exceeded 7,000 *mu*. A 2003 government report on Luntai County notes that the annual net income per farmer/herder in the Caohu District for 2003 reached 3,037 yuan, an increase of 1,872 yuan per head compared to 2001.

The following is the result of a statistical survey of 142 migrant households (624 persons) that I conducted during the first and second stages of migration. The surveyed households owned a total of 10,904 head of various types of livestock (66 horses, 531 cattle, 41 donkeys, 64 mules, 459 camels, 6,503 sheep, and 3,240 goats), of which 6,117, or 56.1 percent, were breeding livestock. The number of livestock grown and Layisued in that year was 4,130, representing a reproduction and growth rate of 94 percent. A total of 2,541 livestock were disposed of, with 33 percent being slaughtered. The total number of sold livestock was 1,811, or 71.3 percent. The production of sheep wool was 13,466 kg, of which 11,607 kg, or 86.2 percent, was sold. A total of 1,139 animal skins were produced, of which 949, or 83.3 percent, were sold. The main agricultural products were yellow apricots, white apricots, peaches, and melons. In 71 (77.2 percent) of the surveyed households, at least one member is engaged in manufacturing or other secondary or sideline jobs. This mainly involves the production of pottery items such as bowls, roof tiles, and jars. The workmanship involved in such manufacturing is relatively rough, and the style relatively simple. A large proportion of the people occupied in the production and retail of this pottery are elderly. Another common sideline is the baking and selling of *nan* (a type of flat bread) or operating small restaurants. An additional 59 people from 39 households have taken jobs outside the migrant village, although they are registered as village residents.

According to the survey results (Table 6-1), the proportion of income earned from Animal husbandery by residents of the migrant village fell by 6.4 percent between 2002 and 2003. During the same period, earnings from external work increased by 1.81 percent, while earnings from other income sources increased by 2.98 percent. The proportions of income from agriculture, manufacturing and secondary occupations, gifts from relatives and friends, and property rental, all changed by only small amounts, ranging from 0.02 to 1.4 percent.

Table 6-1. Breakdown of income per head for migrant village residents in 2002 and 2003 (yuan/person)

Indicators	2002	2003
Gross income	2583.63	3967.81
Production costs (costs for feed, etc.)	452.56	692.60
Tax	155.00	238.10
Average net income	1976.07	3037.11

Table 6-2. Total income distribution per head for migrant village resident in 2002 and 2003 (yuan/person)

Income	2002		2003	
Average income	2583.63	100.00%	3967.81	100.00%
Income from livestock	1751.70	67.80%	2436.24	61.40%
Income from growing crops	374.63	14.50%	630.88	15.90%
Income from manufacturing and secondary work	166.10	6.40%	261.88	6.60%
Income from outside employment	74.20	2.90%	186.88	4.71%
Income from gifts from relatives and friends	20.20	0.78%	30.20	0.76%
Income from property rental	15.50	0.60%	25.00	0.63%
Other income	181.30	7.02%	396.73	10.00%

Table 6-3. Breakdown of spending per migrant for 2002 and 2003 (yuan/person)

Indicators	2002		2003	
Average consumption	1183.3	100.0%	1506.4	100.0%
Food expenses	498.2	42.1%	607.1	40.3%
Clothing expenses	281.6	23.8%	394.7	26.2%
Miscellaneous costs (transport, medical insurance, etc.)	297.0	25.1%	363.0	24.1%
Other (education, etc.)	106.5	9.0%	141.6	9.4%

Table 6-2 shows that total income per head for migrant village residents in 2003 increased by 1,384.18 yuan over the figure for 2002, while net income per head increased by 1,061.04 yuan. Other economic indicators also increased markedly over the previous year.

Table 6-3 shows that there was no major changes in spending patterns between 2002 and 2003, although the proportion of spending on food and clothing dropped, while spending on education and medical care increased slightly. In 2003, the ownership level of durable consumer goods rose, and I noted during my survey that the variety and quality of these goods also increased. Of the 142 survey households, only two poor households own black and white televisions – the other 140 households own color TVs. A total of 118 households own washing machines, of which 26 units are two-tank models. A total of 96 households have telephone lines, and expensive products such as VCD players, sewing machines, and mobile phones are steadily appearing within migrant households. There is also a certain improvement in the production methods of the migrants, with the recent purchase of three tractors and the construction of a 700 m² livestock shed.

Social effectiveness

The subjects of the ecological migration project were ethnic minorities living under very poor economic conditions in areas with inconvenient transport systems. Government funding enabled these people were relocate to areas with water and electricity supplies, good transportation infrastructure, and good cultural, educational, and medical facilities. In addition to increasing the incomes of migrants, this initiative served to significantly stimulate the development of urbanization in the migrant destination areas and enhance social stability and ethnic unity.

New information obtained from my survey

(1) Low education levels of migrants (Table 6-4)

Very few migrants have been educated to junior high school or high school level. Of the survey subjects, 15.1 percent had no schooling at all, 42.9 percent attended primary school, 30.8 percent attended junior high school, and 11.4 percent reached high school level, yet not one of the survey participants had a tertiary education. Although it is impossible to make a marked improvement over just a short period of time, with the implementation of the national government's "both bases" plan and improvement in the educational conditions of the migrants, the standard of education amongst migrants is certain to improve.

(2) Living environment of migrant settlements

In a migrant settlement within the Lapa District, over a total area 4 ha, I observed that the village was provided with all necessary living facilities, including a school, mosque, market, bank, small restaurants, medical clinic, public toilets, water pipes, paved roads, and ventilation stations. Transport is extremely convenient, with buses to the center of the county departing every 10 minutes. As explained by Mr. Gao Fucai, a bus driver employed by the Changshun Bus Company in Luntai County, his company owns eight buses to service the migrant village. On every day of the year, buses depart the village every 10 minutes between 9 a.m. and 8 p.m., providing a high level of convenience to residents of the migrant village. To satisfy the requirements for basic education, the local school set up a system based on nine years of compulsory education, providing sound educational opportunities for the school-aged children of migrants.

Table 6-4. Educational standard of Tarim River migrants in 2003 (no. of persons)

Educational standard	Male		Female	
	Rais District	Rap District	Rais District	Rap District
Illiterate or semi-illiterate	12	28	17	37
Lower elementary school	14	32	16	35
Higher elementary school	26	55	29	60
Junior high school	27	60	26	58
Vocational school	4	8	3	6
High school	14	34	7	16
University or other tertiary	0	0	0	0
Total	97	217	98	212

Previously, the migrants lived far removed from the mainstream of the world, in low, damp wooden houses, utilizing the muddy water of the Tarim River. The sandy ground upon which they lived was bumpy; they cut poplars for firewood to cook their meals, and had no supply of water, electricity, or heating. Their mosques were practically in ruin, having been left without repair for many years, and were without imam. To travel to the county for shopping would have taken up to two days each way by either donkey, tractor, or on foot. Children were required to start work grazing livestock or cultivating crops even before they reached school age.

Changes in the lives of migrants following migration can be understood in more detail via transcripts of the interviews that I conducted with them.

Case 1 Improvement in overall conditions following migration
Interviewee: Turhon Dawati (male, 57, previously a farmer at Caohu Farm), head of Lapa District

Q How was your life before migration? And how is it now?

A Before migration we didn't have anything. Our children did not listen to us, and caused us only trouble. Now they go to school. The teachers educate them well, so we feel relieved. Before, getting to a hospital was a big problem. There was only a shabby clinic in the village, with no bed and few medicines, and most of the medicines were almost expired. The "doctors" were lowly-trained hygienists who could not cure anything other than colds. Now we have a hospital, with university-trained doctors, and departments of internal medicine, surgery and pediatrics. The service is very good.

Case 2   Improvement in livelihood following migration
Interviewee: Hamrati (male, 34, previously a farmer at Karqiaqi Farm)

Q What kind of changes have you gone through since migrating?

A We are very happy that the county and department arts and crafts groups come here often to put on performances of singing, dancing, magic, comedy, and acrobatics. The livestock health office is constantly organizing seminars on livestock-*Raising* techniques and prevention and treatment of parasitic infections. They also help us to repair our livestock shed. We learn a lot from them.

Case 3:   Improvement in Education following migration
Interviewee: Ismail (female, 68, previously a farmer at Caohu Town)

Q Were your children going to school before you came here? What about now?

A My two granddaughters – the older one is 10, the younger is 8 – were not going to school before. There was no school to go to. The government has repaired a school for us, and the two can go to school free of charge.

(3) Problems of conflict in the new migrant settlements

The migrants had lived in their original settlements for a long time, working and living together. For this reason, they all developed essentially the same lifestyle and values: the culture of the settlement was well established. The migrants had developed a relatively strong identity and sense of belonging in their former settlement; however, migration not only places these people and their culture involuntarily into an unknown and foreign social environment, it also dissolves to a certain extent the existing web of social relationships and gives rise to "conflicts" with the various other cultures amongst which they now find themselves. Based on this process, a new culture will emerge in the new settlement. There are two possible results of this situation: either the migrants will maintain a relatively independent subculture within the new settlement, or the new and diverse cultural elements will become fused into the migrants' prevailing culture to produce a new settlement culture that is acceptable to all migrants. In the former case, cultural conflicts in the settlement are likely to continue; in the latter case, the conflicts are likely to ease.

The language, lifestyle, customs, and religious beliefs of the Tarim River migrants are different to those of the Han nationality and Hui nationality, and other ethnic minorities living around the new settlement. In addition, psychological factors such as homesickness, imbalance caused by the ups and downs of the migration process, and contradictions of migrant life are making the conflicts between cultures more pronounced. Some migrants feel that they do not need to learn Chinese, that they can get by speaking only Uigur, while other migrants pour extraordinary amounts of time and energy into studying Chinese. Some migrants feel distrust toward support groups that consist of Han Chinese residents. All of these factors limit the integration of the migrants into their new community.

Problems in the process of migration

Problems of the Tarim River migrants

(1) The "three lows" of employment for female Tarim River migrants

In terms of employment, female migrants from the Tarim River score poorly in three areas: work capability, work participation rate, and work output. The working women are mostly engaged in agriculture, the service industry, and the retail sector. According to surveys, the proportion of female migrants engaged in agriculture, forestry, and Animal husbandery Occupies 60.5 percent, which is 4.6 percent higher than the participation of males. In contrast, the proportion of females engaged in occupations that require relatively high levels of knowledge or skill (e.g., medicine) is extremely low. The major reason for this trend is the restrictive effect of the migrants' traditional beliefs. The level of education received by women is relatively low; this in turn restricts their employment opportunities. In addition, as they are unable to undertake heavy labor, their sphere of employment possibilities is limited.

(2) Low rate of breed improvement of major livestock and low level of industrialization of Animal husbandery

Although the number of superior and improved breeds of dairy cattle is constantly increasing in the migrant village, in 2003 the number of such livestock was still low, making up only 28.85 percent of all dairy cattle, while the quality of sheep is clearly deteriorating. Fine-wool sheep and breed-improved sheep make up only 17.4 percent of total sheep – a drop of 1.3 percent since 2002. Furthermore, the level of livestock processing is low, as are processing capabilities. As there exists only a small capacity for high-level processing, there is little value added to livestock products.

(3) Significant limitation of market factors

At present, the development of Animal husbandry in the migrant village is still in its early stages. In other words, it is still in the "self build-up" stage, with efforts concentrated on serving markets within the county or township. As these markets are relatively isolated and limited in terms of consumption, most of the livestock products are consumed by the migrants themselves. This factor clearly limits the development of the migrants' livestock operations. In addition, the sales system of the livestock products is not yet fully established, in that a considerable number of livestock products depend on purchases by unfixed pedlers; this dependency impedes the development of livestock production.

Countermeasures to migrants' problems

(1) Active provision of work-skills training

It is necessary to change the current situation, in which migrants are limited by their low levels of educational, limited livelihood options, and low work skills. Such change can be achieved by providing the migrants with work-skills training. By constantly improving the education levels and work skills of the migrants, as well as liberating them from their ideas, changing their concepts, and encouraging them to place a greater importance on science, the migrants will be equipped with poverty-relief skills and the ability to adapt to the free-market economic system.

(2) Focus on land resources and developing a real estate rental industry

With development of the local economy in Luntai County, a large fluid population entered the center of the county to engage in trade and commerce. In addition to stimulating the economy, these people are also enlivening the real estate rental market in the area around the center of the county. Given that all people entering the county center must pass through the migrant village, the village may be able to develop businesses relating to real estate rental, trade and commerce, restaurant operation, and car repairs, which would undoubtedly lead to greatly increased incomes for the farmers of the village.

(3) Training of a group of people to manage the affairs of the village, and encouraging farmers and herders to work in the fields of industry and commerce

At present, the major bottlenecks in the development of the village economy are difficulties in selling cash crops, forestry products, fruit, and livestock products that have arisen with the restructuring of agriculture and industry and promotion of the urbanization processes. There is an immediate need for the establishment of a group of village managers who can take on the heavy responsibility of promoting sales. Through these people, the surplus labor of the village can be transferred to small cities to engage in jobs within secondary and tertiary industries.

(4) Taking advantage of local characteristics, and developing tourism

To create a tourism economy, ethnic theme parks and agricultural facilities can be developed to provide tourists with the experience of life on a farm. Dadaonan Town would be ideal as a site of rest and recreation for city residents.

(5) Developing feedlot Raising of cattle and sheep, and taking this opportunity to restructure agriculture.

With funding and the introduction of technology, high value-added processing of beef and mutton can be implemented to promote a more rapid development of the feedlot livestock industry.

Lessons and proposals from the Tarim River migration experience

Problems of migration management

The lessons of history teach us that migration is a long-term and extremely difficult process. Cultural conflicts and changes in livelihood brought about by migration cause psychological problems for the migrants as well as other social problems. These factors can then hinder the sustainable development of the migrant community. Therefore, in the management of migration, the following problems require special attention and need to be solved.

(1) The long time required to fuse the culture of the migrants with other local cultures

As people live for a long period of time within a particular and relatively stable culture and tradition, the culture and traditions transmitted through the generations determines their values, inclinations, and lifestyle. As the Tarim River migrants are characterized by a deep-rooted "peasant consciousness" and a "closed mentality," and their ideas and concepts are relatively backward, it is inevitable that as migrants they will cause conflicts with the prevailing culture in their new environment. Furthermore, the migrants are highly conscious of protecting their own culture. The fact that they make little attempt to blend in with the culture of their new environment means that such conflicts will continue for a long time, creating complex problems.

(2) Long-term management of migrants

The placement of migrants, particularly in relocation to other areas, can be considered as a process of re-creating a material and spiritual living environment for them; however, while creating a suitable material environment is easy, creating a suitable spiritual environment is a long and difficult process.

Migrants tend to adapt to a new material environment relatively quickly, as they can see that the houses are newer, larger, brighter, cleaner, and more solid than the ones they lived in previously. In addition, the land is visibly more fertile and flatter, and various kinds of infrastructure and facilities ensure that their new life is far more convenient; thus, they can readily adapt to the new material environment.

Adaptation of the migrants to the non-material aspects of the living environment is a different matter. Firstly, process of social adaptation by the migrants is slowed by the nostalgia that they feel for the human relationships that existed in their old villages and the caution that they feel towards the unknown neighbors in their new settlements. Secondly, as migrants are not highly educated, they have a weak appreciation of laws and tend to express unstable emotions and irrational behavior. In one case, an Uigur migrant got into a fight with a Han Chinese resident of a neighboring village over a verbal misunderstanding; as a result, the migrant was injured. This incident not only affected the image of the migration process, but also adversely affected public peace and hindered progress in the migrants' livelihoods. Thirdly, after arriving in their new settlements, the migrants will live there for a long time. Thus, the management and services provided to the migrants should be sustainable to promote social coordination and ensure that the migrants are able to settle peacefully into their new lives.

In addition, the standard of the migrants' livelihood and work skills is relatively low. For this reason, until the migrants can retrain themselves by developing such skills, the government and community will need to dedicate significant amounts of personnel, material resources, funds, and technical support to aid them in their development. Because of educational and language-based limitations, these tasks cannot be realized over a short period of time. If just a single mistake is made, the migrants' trust in the government will be endangered, triggering hostility toward the government and resulting in instability. Such an eventuality would be extremely damaging to both social and economic development and to the livelihoods of the migrants themselves.

Proposals

In the process of relocating migrants, the government needs to establish the following groups of mechanisms and support policies, as well as further improving the living standards of the migrants.

(1) Establish a migrant aid fund

Over 2,000 farmers and herders from Caohu District left the area where they had lived for generations and moved into an unknown environment, in order to accelerate their economic development. This process of adaptation involved both psychological and social aspects. Non-institutionalized aid and support are totally inadequate for alleviating and surmounting the various hardships and crises arising from this adaptation process. Stable economic support for the migrants can only be provided by establishing institutionalized and long-term support mechanisms.

(2) Raising the consciousness of migrants

It is necessary to provide "support" to ecological migrants and give consideration to the destination of migrants and the economic development of the area. Migrants should be considered as development migrants, and prevented from returning to their original homeland by ensuring a clear improvement in their standard of living following migration.

(3) Establish a system to guarantee a minimum standard of living

A system should be in place to guarantee a minimum standard of living for migrants, principally to ensure aid for the weak and elderly. Such a system would serve to satisfy the basic living necessities of households that lack labor power and the elderly who have nobody to rely on.

(4) Provide specific preferential policies for migrants

The government should provide maximum support for development of the migrants' livelihoods and improve their standard of living through measures such as tax exemptions, electricity concessions, technical support programs, and offering small interest-free or low-interest loans in accordance with the prevailing conditions in the migrant settlement.

(5) Establish educational and cultural support mechanisms

While it is quite possible to provide relief from economic poverty, a lack of knowledge can lead to failure or even ruin for a group of migrants. For the Tarim River migrants, it is essential to establish educational and cultural support mechanisms. One such mechanism would be to provide subsidies or scholarships to the school-age children of migrants to stimulate their desire to study and alleviate the economic burden of schooling on the household. Another mechanism would be to actively provide education to poorly educated adults, as this will give them an opportunity to join the mainstream of the community and to express themselves.

(6) Ensure that the first stage of migration is successful

Taking various measures as necessary, it is important to ensure the success of the first wave of migrants. This will help to reduce the risk of unnecessary spending on personnel, material resources, and funds, and minimize the distress that migrants experience in the process of re-socialization.

(7) Prevent environmental damage at the migrant destination

The ability of the natural environment to restore itself is limited. Through the implementation of the "seal-off and grazing ban" policy in the area where the migrants lived previously, ecological recovery within designated areas is progressing; however, it is also necessary to prevent new forms of destruction to the ecological environment at the site of the new settlement. If a vicious circle of "destruction-relocation-development-destruction" is set in motion, the results would be terrible beyond imagination.

Overall, ecological migration is highly significant in relation to the social and economic development of the middle and lower reaches of the Tarim River. It is not only an important measure in the protection of the ecological environmental, but also a form of support to relieve people in poor districts from poverty and improve the livelihoods and lifestyles of migrating farmers and herders. In addition, ecological migration is beneficial for the advancement of urbanization, restructuring of local industry, and the development of the local economy. Therefore, ecological migration should be considered as a new model, not only along the Tarim River, but also for the whole of Xingjiang or even all of northwestern China, that links restoration and improvement of the ecological improvement with poverty relief and urbanization. The implementation of ecological migration projects, based on scientific research and rational planning, is essential for promoting the development of local economic societies, as well as ensuring that the "west development project" is pursued in harmony with the protection of the ecological environment.

Acknowledgment

I received valuable guidance and advice in the writing and editing of this paper from Prof. Shi Guoqing.

References

Xinjiang Nianjian, 2002. *Xinjiang Yearbook,* 2002. {Chinese}

7

The effectiveness of "ecological migration" in reducing poverty (2)

Lessons from the Implementation of Ecological Migration in Alasha League, Inner Mongolia Autonomous Region

SHI GUOQING

Introduction

Alasha League is located in the western part of the Inner Mongolia (Nei Menggu) Autonomous Region. The total area of the league is 270,000 km², representing 22.82 percent of the total area of the autonomous region. Grasslands cover 187,400 km², or 69.41 percent of the league. Of this area, usable grasslands total 101,500 km². In theory, this area of grassland should support the cultivation of 1.904 million head of livestock. The forest resources of the league consist mainly of natural forests. The total area of forest is 882,900 ha, with 22,400 ha of arable land. Mountains extend around the periphery of the league, while the interior comprises the Gobi (Gebi) Desert. Other deserts, such as the Bdaiinjaran (Badanjilin), Ulaanbuh (Ulanbuhe) and Tengger (Tenggeli), extend throughout the league. These deserts consist of a mixture of lakes, hills, flatlands, and sand dunes. The climate of these areas is a typical temperate continental monsoon climate, characterized by extremely low precipitation, strong winds, and large volumes of sand and dust. A distinctive feature of the area is windstorms that occur frequently during winter and spring. Along with drought and sandstorms, these windstorms are the major limiting factor to agricultural and livestock production (a form of livelihood that combines the cultivation of feed crops and medicinal plants with the raising of livestock) in these areas.

Alasha was once a region of beautiful grazing lands, rich in water and grass. The Ejene (Ejina) Oasis, *Suosuo* Woodlands, and secondary forest of the Helan Mountains formed an ecological barrier within the Alasha region, protecting the development of agricultural and livestock production not only within the region, but as far as the Hexi Corridor in Gansu

Province, the Ningxia Plains, and the Ordos (Hetao) Plains of Inner Mongolia; however, the Ejene Oasis in Alasha League is on the verge of vanishing because of several factors that have developed over the past 40 years, including the drying of rivers due to falling precipitation, insufficient protection of the Ejene Grasslands, and projects such as irrational development. The Ejene Natural Forest has suffered severe damage, and the issue of desertification is becoming increasingly prominent, as sandstorms are a common occurrence and the ecological environment is deteriorating rapidly. At present, a total area of 222,300 km^2, or 82.3 percent of the total area of the league, has become desertified. This area accounts for 25.6 percent of all the desertified area of the Inner Mongolia Autonomous Region. At present, desertification is expanding at a rate of 1,000 km^2 per year. The problem of sandstorms, which are indicators of environmental deterioration, is becoming progressively more acute in the Alasha region. As well as occurring with increasing frequency, the damage caused by sandstorms is extending further each year, to as far as northwestern and northern China. The sandstorms that occurred in the seven straight years from 1993 to 1999 caused direct economic losses of up to 500 million yuan. In 2000, 21 sandstorms occurred in China, while in 2001 there were 30. Of the 21 sandstorms that occurred in China in 2000, eight originated in Alasha. Thus, specialists consider Alasha to be one of the major sources of sandstorms in China.

While Mongolians are the principal ethnic group in Alasha League, the league is home to significant populations of various other ethnicities, as well as many Han Chinese. The total population of the league is 182,800, of which 57,200 live in rural areas. There are three banners within the jurisdiction of Alasha League: Alasha Left Banner, Alasha Right Banner, and Ejene Banner, which as of 2000 together consisted of 12 towns, 32 sums (equivalent to a township), and 228 Gachaas (equivalent to a village). The average net income per farmer or herder was 2,101 yuan in 1998, 2,284 yuan in 1999, and 2,434 yuan in 2002. These incomes are within the middle-upper bracket of incomes for the autonomous region, and are also high in absolute terms; however, the standard of living of the farmers or herders can be better understood by the following description. Herders live dispersed over wide areas. Most have no electricity and no means of communication. The costs of transport, education, basic necessities, and food are high; the cost of living is much higher than in other areas of the Inner Mongolia Autonomous Region. Environmental conditions in Alasha League are very poor, and infrastructure such as transport, communications, culture, education, and hygiene are poorly established. Thus, it is difficult to improve the livelihood and living conditions of the nearly 25,000 herders in the entire league. In addition, natural conditions continue to deteriorate: desertification is becoming increasingly severe and grass resources have completely dried up. Natural disasters occur frequently, and drought and sandstorms cause enormous damage to farming and the raising of livestock. Each year, losses suffered by farmers or herders amount to approximately 20 million yuan: an average loss of 500 yuan per farmer or herder. Thus, the life of farmers and herders is unstable. It seems that the destruction of grasslands in Alasha League is directly connected to ecological degeneration, the scattering of herder households, and a pattern of grazing that depends heavily on nature. It would be extremely difficult to solve the environmental and economic problems facing the herders in Alasha League without relocating the herders to break their dependency on the natural grasslands. Judging by the extent of poverty and the natural conditions in Alasha League, it is necessary to migrate at least 18,000 more herders through the policy of "ecological migration." This is the only way to significantly improve the livelihoods and living conditions of the herders. Of these 18,000 migrants, 6,000 herders can be classified as poor, while the other 12,000 are affected by insecure incomes and live on the borderline of poverty.

Ecological migration and poverty relief migration to other areas is very important for ecological environment protection in Alasha League. This policy is the key to relieving the poverty faced by herders in Alasha League and is an important measure in implementing the "National 87 Poverty Relief Achievement Plans" and "Ecological Migration Strategy."

Ecological migration projects in Alasha League and their implementation

I. The history of Alasha League

The condition of the livestock-raising industry in Alasha League is very fragile. Since the 1960s, a chronic shortage of fodder has meant that large amounts of feed and fodder have had to be brought in from outside the league to protect livestock in the event of natural disasters. That was a rather passive means of protecting livestock against disasters; consequently, in the mid 1970s the Alasha League began to develop and construct livestock feed "stations." A series of small- to medium-sized feed stations were built at sites such as Yaoba, Chahar (Chahaer), Luanjingtan, and the Ejene Riverine. Attracted to the feed stations, large numbers of people began moving to these areas. By the late 1980s, there was already 10,700 ha of land under cultivation, and at this time the government proposed a "strategy shift" with regard to ecological migration. The government began implementing a voluntary migration program whereby it assisted herders to relocate and engage in farming at areas already developed. By supplying fodder and feed in livestock raising areas, this support eased the grass shortage and led to increased income for farmers and herders. Poverty-relief activities also followed in the same context, and coordination was implemented to achieve concentrated poverty relief via the migration and relocation of herders.

II. The necessity of ecological migration

Large numbers of people in Alasha League live in areas of very poor natural conditions; ecological conditions have deteriorated to the point where the inhabitants cannot sustain a living. Grassland resources are limited, and overgrazing and excessive livestock numbers pose serious problems. According to the estimat, the grasslands of Alasha League can sustain a total of 1.5 million head of livestock, however, there are already 2 million livestock in the league. If grasslands deteriorate further, they will provide insufficient basis for sustaining a livelihood, and herders will be vulnerable to poverty.

Alasha League extends over a vast area, and people in livestock-raising areas live scattered far apart. As a result, substantial investment is required to construct various kinds of infrastructure. Many livestock-raising areas thus lack the necessary conditions for development. For example, where fluorine and arsenic in groundwater exceeds standard levels, it is difficult to secure drinking water for people and livestock. This in turn makes it difficult to improve economic conditions.

Relocation via ecological migration is necessary for the sustainable development of the Alasha League economy. The league has a sparse population and a vast area of land. Basic conditions, such as the large number of resource production sites, are relatively good, so ecological migration is easily accommodated. Sustainable development of the Alasha League economy should be realized by relocating subsistence herders to these resource production areas, implementing poverty relief through development, and restoring grassland vegetation via grazing suspensions.

III. Favorable conditions provided by ecological migration

(1) The current development of ecological migration represents a rare historical opportunity.

Ecological migration accords well with the policy of the "West Development Project," economic restructuring, urbanization development trends, and the pressing desire of society to "become affluent."

(2) The "West Development Project" provides an excellent opportunity for the migration of farmers and herders.

The most important elements of the "West Development Project" are ecological protection and poverty relief. As Alasha League is an important area for the improved protection of the ecological environment, it is included in the list for the first stage of the implementation of the migration plan. Thus, intensive support will be provided in the field of poverty relief for migrants and protection of the ecological environment.

(3) Migration and relocation are widely accepted by leaders and people alike.

Severe droughts in recent years have caused the death of large numbers of livestock, serious deterioration of grasslands, and the frequent occurrence of sandstorms. In the face of such conditions, local leaders and inhabitants have reached a consensus that migration and relocation are historically inevitable.

(4) Migration and relocation in Alasha League has achieved positive results and experiences.

In the several years that migration activities were first implemented, the lifestyles and livelihoods of farmers and herders who have migrated have changed dramatically. The problem of poverty has essentially been solved, and the local governments who participated at each level have accumulated abundant experience that constitutes a foundation for successive migration projects.

(5) Migration conforms to the wishes of farmers and herders and is an important means of improving their livelihood and lives

The ability of herders to subsist on grazing in the deserts and Gobi has been steadily eroded. In recent decades, the number of livestock has decreased, and it is difficult for herders who rely on traditional livestock management methods to improve their standard of living. The most pressing desire of herders is to escape poverty immediately. This ideal can only be realized by relocating herders from areas of deteriorating subsistence conditions and "national ecological environment protection construction areas" to regions of relatively good water and soil conditions and by promoting intensive development.

(6) Financial guarantees have gradually become possible with economic development

Because of the relatively large budgets involved, in past times such migration programs would have collapsed at the planning stage; however, with the considerable development of the regional economy of Alasha League, the level of investment in the "national ecological environment construction fund" in Alasha League has increased manifold; sufficient funds are now available for migration programs.

IV. Guiding concepts

Utilizing the excellent opportunity presented by the "West Development Project," the league, committee, and other administrative offices are to guide the ecological migration project based on a strategy of "moderately compact and relatively concentrated relocation and development." These parties aim to protect and restore the ecological environment in Alasha League, starting with improving the subsistence conditions of herders and increasing their incomes. By focusing on the distribution of livestock activities and industrial structure, they are to rationally locate livestock-raising areas. After relocating farmers and herders from areas subject to the project and undertaking industrial restructuring, they are to implement grazing bans and restore vegetation in areas that are currently wasteland. Furthermore, in line with the slogan "efficiently develop and utilize small areas of land, and protect and restore large areas of land," they are to realize sustainable development of the ecological environment and the economy of livestock-herding areas. By formulating rational policies and applying modern science and technology, they are to proceed with afforestation of the lands vacated by migrants and ensure a secure, prosperous life for migrants in migration destinations.

V. The overall concept of ecological migration

(1) Connection between ecological migration and ecological construction.

The policy of ecological migration seeks to conduct ecological relocation of herder households currently living in deep inland desert areas and the Gobi who have lost their subsistence conditions, as well as herder households in "Helan Mountains National Nature Reserve" and "*Suosuo* Woodlands Reserve" and solve their basic livelihood problems. In addition, the policy aims to conduct forest and grassland restoration and ecological management in the areas vacated by migrants to enable vegetation to regenerate.

(2) Combining ecological migration and industrial restructuring in livestock-raising areas.

The project aims to carefully and systematically evaluate historical experiences and compel herders to abandon their backward concept of traditional grazing-style livestock raising, which in the process of deriving value sacrifices the grasslands. This traditional approach aims only at increasing the number of livestock, without concern for quality, and depending only on nature. In addition, herders are encouraged to implement regional development based on science and technology. The policy also involves the restructuring of local industry, the formulation of various preferential policies, and the encouragement and cooperation of herders in their migration and relocation, as well as their engagement in agriculture and feedlot livestock production. At the same time, the policy seeks to diffuse scientific education and foster values of caring for grasslands and protecting the environment amongst herders, as well as promoting the sustainable development of livestock raising.

(3) Connecting ecological migration with the implementation of the "two rights, one system" policy in grassland areas.

The introduction of a user-pays contract system in terms of the ownership and utilization rights of grasslands and pasture has had an important effect on improving grassland environments; that has been verified by practice. Along with the implementation of ecological migration, accelerating the execution of the "two rights, one system" policy for grasslands and pasture in the project area ensures a more rapid recovery of natural grasslands.

(4) Connecting total migration and dispersed migration.

Migration is implemented principally on the basis of relocating an entire Gachaas. The placement of individual households should follow an overall arrangement. This approach ensures better organization and guidance of migration projects and results in better arrangements for the new livelihood and lifestyles of migrants.

(5) Connecting ecological migration to the abolition and amalgamation of Gachaas and Sums and with small city construction projects.

If all projects are interlinked, ecological migration is convenient not only for concentrating organization and unifying guidance, but also for avoiding overlapping construction projects and unnecessary spending and waste. This also enables ecological migrants to completely break away from their old lifestyle, which is lonely, indifferent to culture, primitively closed, and backward.

VI. *Basic methods*

(1) Targets

Between 2001 and 2005, a total of 24,467 people were to be relocated. In principle, large-scale land developments would not be undertaken; instead, previously developed land would be rationalized (or small amounts of land would be appropriately coordinated and developed). A total of 1,130 ha of agricultural land and 3.626 million ha of grassland were to be abandoned by the departing herders. The income of relocated farmers and herders would be increased to "breathing space" level and the progressive ecological deterioration of the area in which they resettled would be arrested.

(2) Selection of migration subjects

The main subjects of the migration policy are populations of herders who possess livestock and occupy grasslands. In poor areas, areas of severe ecological degradation, and areas in which it is difficult to secure drinking water for people and livestock, it is necessary for the entire population to migrate. After the abolition and amalgamation of Sums, surplus inhabitants are to be intensively assigned to jobs in secondary and tertiary industry in cities and economic zones. In principle, the decisions concerning which households are to migrate will be in accordance with the wishes of the household. At the time of actual implementation of ecological migration, migration subjects are selected according to the following principles: 1) Poor Gachaas and herder households with an income per head below a specified poverty level ("*wenbao* line"); 2) Herder households living in the desert or Gobi where conditions make it difficult or impossible to survive 3) Herder households who have difficulty securing drinking water for people or livestock, or whose drinking water contains fluorine levels far in excess of health standards; and 4) Herder households living within the "Helan Mountains National Nature Reserve."

(3) Migration destinations

According to the process of industrial restructuring in Alasha League, migration locations need to involve multiple routes on multiple levels. There are three main approaches. One is to move away to different area and engage in agriculture or livestock raising; the second is to remain in the same area but to undertake a change in livelihood, engaging in specialized agricultural production; the third is to remain in the same area and undertake employment in the secondary or tertiary industries. The relocation destinations are principally irrigation districts on the Yellow River, including Luanjingtan, Bayanmod (Bayinmaodao), Bayanmuren (Bayinmuren), and Xitan, or the Ejene Oasis Irrigation District within the lower reaches of the Heihe River, which is already concluded in the national plan, as well as towns such as Jirantai (Jilantai) and Usutu (Wusutu). Land within migration destinations is used in a rational way, and emphasis is placed on measures that reinforce and reorganize existing agricultural land. Science-based agriculture and livestock raising are to be linked with cultivation, livestock raising, and processing; agriculture, forestry, and the livestock industry; and agriculture, industry, and commerce. The placement of migrants in jobs within the secondary and tertiary industries is to be conducted under government guidance with private sector funding. Appropriate arrangements should be implemented for farmers and herders who opt for migrating, and the government needs to have unified control over the land and plains in areas vacated by migrants to prevent migrants from returning. While deciding on important matters such as migration and the relocation destinations of farmers and herders, the national government should give sufficient consideration to the circumstances of the three banners in the league, listen to the opinions of local governments, and guarantee the rationality and implementation of the migration plan.

The basic criteria for selecting migration destinations in the case of ecological migration within Alasha League are: 1) Convenient transport; 2) Flat land; 3) Relatively good soil conditions; and 4) A minimum quantity of water resources. With providing these basic conditions, areas of relatively sound natural conditions in each banner are to be selected as migration destinations.

The ecological migration and poverty-relief migration projects in Alasha League involve three banners and 36 townships, towns and sums. The total number of people to be migrated or relocated over five years is 24,467, from a total of 6,224 households. Of these migrants, 15,538 people from 3,979 farmer or herder households are to move to ecological districts such as Luanjingtan and Chenjiajing to engage in ecological farming: a combination of agriculture and livestock raising. Another 8,149 people from 2,041 farmer or herder households are to move to urban areas such as Bayangol (Bayanggaole) Town, the Usutu Development District, and Lujiajing district to take jobs in secondary and tertiary industries such as the service industry and minerals development. A total of 543 people from 140 herder households are to be relocated to Hoshochi (Hushu) Gachaa in Tamsu (Tamusu) Town to engage in planting *Suosuo* forest, cultivating *Cistanche deserticola* YC Ma (Roucongrong), and breeding camel and white cashmere goats. A further 240 persons from 60 herder households are to be relocated to a salt lake agricultural development district and a tourism development district.

(4) Means of livelihoods

1) Construction in the existing ecological agricultural development district. This is in addition to conducting careful cultivation with stable land plots, an increasing use of science and technology, implementing water-saving agriculture, and increasing the farmer and herder population.
2) Development of specialized agriculture and livestock raising in areas of desert and salt lakes, and compelling herders in districts with certain conditions to change their livelihood.
3) Development of tertiary industries, mainly tourism, services, and intermediary agents, and finding new work places for migrants.
4) Reducing grazing-based livestock raising and developing feedlot-based livestock raising, plantations and orchards, and the cultivation and processing of grazing grass. Accordingly, these policies will result in increased income for farmers and herders and improve the general ecological state of the agricultural industry.

(5) Compensation policies

The national government provides a certain quantity of grant-in-aid for the construction of housing and livestock sheds for herder households who are migrating to farming areas to engage in farming. As well as relocation costs and grants-in-aid for living support, the national government provides migrants with the compensation of arable land (completely furnished with water supply, electricity, and protective forest) of the same value as that vacated, including land, grassland, and production facilities. For both farming areas and livestock-raising areas, the government does not provide compensation for usable livelihood and living materials.

For relocated herder households who take employment in secondary and tertiary industries, the national government provides compensation for wells, livestock sheds, silage, and fencing for grassland and grazing land. In addition, the national government provides a fixed amount of grant-in-aid for relocating, grants-in-aid for housing and supporting the expenses of the new livelihood, and favorable policies regarding land for business use. The level of compensating for ethnic minority farmers and herders needs to be improved appropriately, and grain is to be provided to such migrant households for several years following their relocation.

The system implemented for ecological migration within Alasha League is described later in this chapter.

Systematic relocation of migrants

Case one: study of migration destinations: Luanjingtan Irrigation District

The irrigation district in Luanjingtan is located within Jargalsaihan (Jiaergalesaihan) Town, southern Alasha Left Banner. The irrigation district is well connected to other areas by transport and communications links: access to surrounding areas is convenient. The planned area of the irrigation district is 16,400 ha, with a development area of 11,500 ha. The geographic features of the irrigation district are dominantly open, flat topography with tanso calcium soil and swamp calcium soil. As the conditions of water, light, heat, and soil are suitable for the growth of various types of crops, it's an ideal area for agricultural cultivation.

Construction of the "Yellow River pumped hydropower program" in Luanjingtan began in September 1991 and operation began in 1994, thereby enabling large-scale and comprehensive agricultural development. The area of land already developed in the irrigation district is 7,500 ha, of which 6,800 ha is arable land and 870 ha is forest. Now that water channels, farmland, roads, and forests have already been established in the district, farming and livestock raising can be developed simultaneously by the combined use of farmland, grazing land, forest, and grassland. In the construction of water channels, anti-leak measures were applied at three levels: the trunk, branches, and source of the water channel. At present, about 8,000 subsistence herders from 1,800 households have relocated from 22 sums and 64 Gachaas. In 1999, a total of 5,300 ha of agricultural crops were planted in the irrigation district, yielding a gross crop production of 16.2 million kg. Average yield was 6,975 kg/ha, and net income per head was 1,500 yuan. Overall, the irrigation district is promoting the adaptation of new technology for farming and livestock raising. The entire process from sowing seed to harvesting is now mechanized, and more than 60 percent of migrant households own small- to medium-sized tractors. Utilization of the straw and stems of agricultural crops for livestock raising is spreading rapidly, and the scale of livestock raising in the district stands at about five sheep per household and one pig per person. Although in its initial stage, a sound ecological cycling system is now taking shape, in which forest protects farmland, farming feeds livestock, and manure aids farming.

Case two: ecological migration policy system

(1) Construction of living and livelihood facilities

The ecological migration policy provides for the construction of a house of a specified size free of charge to migrants, and guarantees 1.3 to 2 ha of arable land as well as a water supply and electricity for each household. There are some differences in such provisions between different migration destination areas, depending on prevailing conditions. For example, in the Yellow River Irrigation District in Luanjingtan, water supply and electricity are relatively expensive, but a large area of land is provided to migrants – up to 2 ha. In contrast, within areas with good water supply and land conditions and relatively little investment funding, such as Xiangxitan and Manshuitan, migrants are provided with just 1 to 1.3 ha of land per household. Rights to the land are valid for 30 years. In addition, livelihood facilities such as feedlot sheds, silos, and greenhouses are also built for migrants. Minority ethnic households are provided with 30 percent more grant-in-aid than other migrant households.

(2) Tax revenue policy

For five years from the year of migration, migrants are exempted from paying agricultural tax and special product tax at the migrant destination. Migrants are also entitled to benefits through other preferential poverty-relief policies. For example, migrants with children of compulsory schooling age who face financial difficulty are granted exemptions from paying tuition fees and miscellaneous educational costs for two years.

(3) Land use rights for the grasslands previously owned by the migrants

To ensure smooth progress in restoring the ecology in the areas vacated by migrants, the abandoned grasslands are to be uniformly managed by the government. The government is also to ensure that migrants do not return to their deteriorated grasslands to conduct grazing and thus cause further damage.

(4) Management of family registers

As it is necessary to record statistics of migrant farmers and herders on a household basis, people are to be appointed to manage statistical records. While it is necessary to give sufficient consideration to the wishes of farmers and herders, they are to be encouraged to enter secondary and tertiary industries. Registration is to be conducted under the agreement of three parties: the banner government, the sum government, and the farmers and herders themselves. No one is permitted to change registration freely. When there is any change to a family register, the three parties must confirm the changes via consultation. Copies of the family registers of migrants should be kept at police stations in migration destination areas. Previous family registers held at the pre-migration site are to be erased and the registers are to be changed to "city registers."

(5) Poverty relief policy

Poverty relief projects across the entire league are focused on migrant households. Thus, funds for poverty relief are to be used mainly to alleviate poverty within the poorer migrant households. In addition, "comprehensive agricultural development investment" is to be urgently implemented. This investment, along with poverty relief funds, is to be used for the workplace protection of migrants and also for relative poverty relief between government leaders and poorer households.

(6) Plan and arrangement

In principle, migrants are to secure a means of livelihood prior to migration. The government is to ensure that only those people who have fulfilled the conditions for migration will migrate each year, to ensure that the overall migration plan is completed. The government is also to guarantee that the income level of the herders who migrate does not fall following migration and that their standard of living improves. That is to be accomplished via a process that integrates migration, settling down, increasing income, ecological restoration, and economic effects.

(7) Construction at the migration origin

Construction at the migration origin is to be conducted under the overall migration plan with regard to the protection and construction of ecological environments in Alasha League. The entire area it to be divided into smaller areas according to location, and each of these areas is to be managed separately. The main activity will be the protection of areas via the discontinuation of grazing. In addition, in important areas, seeding is to be undertaken using airplanes, and biological and technological measures of land management are to be adopted. In this way, the areas where grazing has been discontinued because of migration or changes in the livelihood of inhabitants, which will serve as important models of land management. High-quality model areas are to be created by appropriate irrigation, using the irrigation facilities employed by the former inhabitants; this will accelerate the restoration and ecological management processes. In addition, necessary rules and systems are to be implemented to strictly ban all forms of grazing and prohibit the return of migrants to their former grasslands to resume livestock raising. Depending on the physical conditions of the area of migration origin, the natural restoration of natural vegetation is in principle to be promoted mainly by protecting the land via ongoing bans of livestock raising in these areas. There are three major programs designed to achieve these aims: the program of protection for the three great natural forests, the natural secondary forest of the Helan Mountains, the *Suosuo* forest that extends throughout the league, and the Ejene poplar forests; a program to construct ecological barriers; and a program to construct the Ejene Oasis Corridor.

(8) Construction of migration destination areas

Various projects, such as the relocation of migrants, making arrangements for their living and livelihood, and the provision and management of social services should be implemented according to the principle that migrants are relocated systematically and in stages. It is expected that ecological migrants will inevitably confront a certain amount of difficulty with regards to lifestyle, livelihood, and generating income during the first year of migration. To help migrants engage in their new livelihood with a degree of security, the national government and each level of local government need to support migrants in overcoming their problems by providing appropriate support in terms of relocation expenses, the construction of housing, and living expenses. The arrangement for migrants' livelihoods is the basis for the overall aim of realizing the ecological migration program. To achieve this, it is necessary to utilize the advantageous features of resources in the migration destination. As each migrant has a different level of technical and cultural proficiency, the strengths of individuals should be utilized in addition to considering the suitability of land in the migration destination when deciding on suitable migration destinations and selecting suitable livelihoods. For example, if an area of land is suitable for farming, then it should be used for farming; if a piece of land is suitable for livestock raising, then it should be used for livestock raising. When making arrangements for the livelihoods of migrants, in addition to considering and balancing present living and economic conditions, it is also important to look into the future and make use of advantageous features of local areas to ensure local economies are able to develop in a sustainable manner.

The majority of migrants relocated from areas which were purely livestock-raising areas. They therefore have little understanding of farming in agricultural areas and other occupations. Thus, following migration, it is necessary to provide them with a certain level of livelihood skills by training in essential, practical occupational skills and improving their familiarity with science and technology. As the migration program involves a shift from living loosely in scattered locations over a wide area to living in concentrated settlements, once the migrants have settled down it is necessary to reinforce their education in different fields such as culture, healthcare and hygiene, science and technology, ecological protection, and policy and law. Active guidance should also be provided to motivate migrants to make efforts to create "civilized" townships and towns, and new "civilized villages." At the same time, the level of social service and management should be enhanced to realize the success of the ecological migration program in terms of both materialistic and spiritual aspects of "civilization."

(9) Sources of funding

A pluralistic investment system has been established as a source of funds for the migration. Four methods have been adopted to procure funds from a wide variety of sources. In the first method, funding is provided by the national government. According to related policies, the national government provides 4,500 yuan to each migrant household when implementing the migration program. In the second method, funds are provided from the resources of local governments. The two levels of local government, league and banner, divert a portion of the funds budgeted for purposes such as agricultural development, poverty relief, and job creation to the migration program. In addition, various sections of government, such as agriculture, livestock raising, forestry, irrigation, and electricity, provide a portion of their construction funds for the development of migrants by organizing projects and asking the migration program to support the projects. The agricultural electricity section is able to construct electric power facilities in the migration destination areas, while the irrigation section is able to undertake irrigation work in migration destination areas, the forestry section can be used to improve the protection of forests in migration destination areas, the livestock-raising section can arrange for the construction of livestock sheds and silos, and the agriculture section is able to construct livelihood facilities such as greenhouses for farmers. The "poverty relief secretariat" is able to dedicate specific funds for use in supporting the livelihood and living conditions of poor migrants, while the "Public Welfare Bureau" is responsible for managing the emergency relief fund to assist the victims of natural disasters, and to secure grants-in-aid to compensate for difficulties faced by migrant households. The agriculture, forestry, and livestock-raising sections can each link with the migration project via approved construction projects and thereby invest money in the migration destination to increase funds for the migration project. The third method of raising funds is to seek funding from private companies and other aid funds. This can be done by encouraging large companies within Alasha League and companies in various fields who are engaged in environmental projects in the league to actively participate as investors in migration projects. Funds can be secured from companies in exchange for land, and the companies can be encouraged to conduct ecological environment restoration in both migration destination and migration origin areas. The fourth method involves funding of the migration by the migrants themselves. Assuming that the migrants are able to bear the financial burden, each migrant household is asked to provide a small sum of money that is then used for the implementation of the migration policy.

(10) Social service system

Each level of government and all sections should set up service stations at the migration destination; for example, agricultural and forestry service stations, veterinary stations, irrigation stations, grassland management stations, local medical care stations, and culture stations. These stations can help to solve the problems of the farmers and herders related to livelihood and living. They can also provide a comprehensive range of services to the migrants, including livelihood assistance, living materials and technical services, help to promote awareness of science and technology, and consultations and training for the migrants. Through these measures, the stations can guarantee to solve the real problems faced by the migrant households. Each section needs to select suitable technicians to work at the service stations. The necessary qualities for such technicians are a strong sense of responsibility, a high level of specialized skills, and the strength to endure hardship and live for an extended period within the migrant area. If these technicians can be selected from well educated and appropriately skilled migrants, the employment of such people will serve to both secure a livelihood for migrant households and overcome the difficulty of finding government employees that wish to work in rural areas. Each service station can hold regular technical training courses to provide experience, information, and special skills that enable farmers and herders to increase their income via the adoption of scientific methods.

Migrants are to be actively guided to engage in employment within secondary and tertiary industries. Migrants who work in businesses such as transport, restaurants, and distribution can stimulate the market and bring benefits to both households engaged in farming or livestock raising and those engaged in employment within the two fields.

Prior to the implementation of full-scale migration, it is advisable to initially relocate a number of migrants who are relatively favorably placed for migration. After providing these initial migrants with support and guidance in all areas, they can be used as model migrants for the migration process. The example of these first migrants will ensure a smoother migration for other farmers or herders, who can thus enjoy higher incomes better.

VII. Relocation models for ecological migration

Excellent results can be achieved by relocating ecological migrants in Alasha League in a variety of forms appropriate for each area, as well as by adopting different relocation models, such as farming, livestock raising, special feedlot raising, and employment within secondary and tertiary industries, based on the preferences of migrants and prevailing economic conditions. Please refer to the following descriptions for typical models to each case.

Ecological migration relocation model 1: relocation to farming and livestock raising

In migration destinations with favorable conditions in terms of nature, climate, and soil, high-efficiency farming with water-saving-type irrigation can be actively implemented by utilizing previously developed land and irrigation resources, as well as by improving irrigation, electricity, and other secondary facilities through rational planning. Based on the relevant ecological migration regulations set by the government, and with reference to the prevailing situation within Alasha League, migrants are to be provided with approximately 0.5 ha of land to be cultivated. Basic farm crops are corn and wheat, but effort can also be invested into the development of feedlot livestock raising by making use of the abundant feed resources generated by farming. In addition, intensive production and management skills can be promoted amongst the farmers and herders by training in science and technology and encouraging a shift away from the concept of traditional natural grazing as the only form of livestock raising. The number of livestock engaged in feedlot raising can be increased by the cultivation of high-quality and high-yielding grass and feed crops and implementing various test models for the feedlot raising of cattle, sheep, and pigs. It is important to improve production management, improve profit levels, and gradually increase the production of local specialties. There is also a need to improve both farming and livestock-raising practices, and accordingly quickly solve the problems of migrant households concerned with their livelihood and living conditions and establish a sound base for future development.

Migrant households in districts where both farming and livestock raising are pursued can select appropriate production methods based on objective conditions such as land and irrigation facilities and individual production skills. They can develop water-saving irrigation farming by planting appropriate proportions of crops such as corn, wheat, and potato. At the same time, migrants can cultivate high-quality grass and feed crops such as alfalfa, Shadawang, and Yinghong corn. The migrants can also conduct feedlot livestock raising of dairy cattle, beef cattle, cashmere goats, Xiaoweihan sheep, pigs, and chickens, by making use of the straw and stalks of farm crops or other farming by-products. They can also actively develop feedlot livestock raising. By combining farming and livestock raising, the migrants can not only satisfy their basic subsistence needs but also assist in the economic development of the area.

Farming production skills involve improvements in seed varieties, water-efficient irrigation, self-supplied organic fertilizer, small-scale farm machinery, and cultivation methods that consider farmland conservation.

The relocation methods for households engaged in farming and livestock raising are as follows:

(1) Each household is provided with 1.3 to 2 ha of land for crop production. The migrants cooperate to prepare protected forest, irrigation, an electricity supply, and land.

(2) Direct investment in migration (grant-in-aid to migrant households) involves: (a) for housing construction: $30\,m^2$/household × 350 yuan/m^2 = 10,500 yuan; (b) for feedlot shed construction: 50 m^2/household × 100 yuan/m^2 = 5,000 yuan; (c) for relocation expenses: 1,000 yuan/household; (d) for income stability: 1,000 RMB/person × 4 persons/household = 4,000 yuan. Thus, the total grant-in-aid per household amounts to 20,500.

Ecological migration relocation model 2: relocation to farming and livestock raising of specialized products

Many local livestock species in Alasha League are very useful and possess special features. Special local livestock breeds include white cashmere goats and Alasha two-humped camels. Alasha League is also a major area of production for Cistanche deserticola YC Ma. In the past, people harvested all the wild Cistanche deserticola YC Ma, causing severe damage to natural Suosuo (a type of shrub) forests. Skills for the cultivation of Cistanche deserticola YC Ma have since matured. Some of the farmers or herders in areas of Suosuo forest are subject to "grazing discontinuation for forest restoration" and are therefore compelled to plant Suosuo shrubs. They are then allowed to graft Cistanche deserticola YC Ma onto the Suosuo shrubs. In addition, Alasha League is home to many valuable medicinal plants, including ephedra, licorice, and bitter bean grass, as well as special arid-region edible plants such as desert mugwort, desert rice, and desert onions.

The relocation methods for households engaged in farming and livestock raising of specialized products throughout the league are as follows:

(1) Migrant households engaged in the farming and livestock raising of specialized products are required to sell the land they vacate to ensure that they are no longer able to engage in grazing. The number of live-

stock that each household can maintain should not exceed 10, and the migrants should not be allowed to graze their animals on grasslands.
(2) Each household is provided with 100 ha of Suosuo forest as production land, as well as irrigation facilities and wells. Grazing bans are imposed on all the grassland vacated by the migrants, and this grassland is to be managed systematically by the government.
(3) The farmers and herders are provided with skills training in planting Suosuo shrubs and grafting and managing Cistanche deserticola YC Ma.
(4) Each household is provided with 1,000 yuan of production materials as a grant-in-aid.
(5) Each household is provided with a grant-in-aid of 6,000 yuan for livelihood assistance. Grain and basic necessities are also provided for a period of three years.
(6) Each household is provided with a grant-in-aid of 18,000 yuan (4,500 yuan per person) to help cover the expenses of commencing production. Thus, a total of 24,000 yuan in grants-in-aid is paid to each household.

Ecological migration relocation model 3: relocation to occupations within secondary and tertiary industries

The entire area of Alasha League is relatively rich in mineral resources. Mining ventures include the Zonbil (Zongbieli) Coal Mine and Jirantai (Jilantai) Salt Mine in Alasha Left Banner, and the Gashontata (Nurigaisumugashundada) Multimetal Mine in Nuurgei (Rurigai) Sum, within Alasha Right Banner. At Gashontata, within Nuurgei Sum in Alasha Right Banner, projects for the mining of metals including gold, cobalt, and gallium have been confirmed, as well as magnetic iron ore mines. The first stage of construction of the Gashontata iron ore mining and dressing project is already complete, with a production capacity of 400,000 tons of iron ore per year. Eventually, these mines are expected to produce 1.2 million tons per year. The development of mineral resources is also expected to stimulate the development of transportation, trade, restaurants, service industries, and the entertainment industry. If a rational industrial structure is established, farmers and herders who move to the area can renounce livestock raising completely. By moving out of the grasslands where they currently live, the ecology of the area can be protected. The development of the mining industry in this area can directly employ 240 people for mining and dressing, and more than 300 in transportation, trades, restaurants, service industries, and entertainment.

The method of shifting migrants to occupations within secondary and tertiary industries is as follows:

(1) The family registers of migrant households engaged in secondary and tertiary industries are altered to non-agricultural family registers.
(2) Such migrants are provided with free land for establishing a business in mining areas or along national highways. The area of this land should be no less than 600 m^2 per household.
(3) Direct investment is to be used for migration expenses (grants-in-aid to be provided to migrant households). For transportation expenses: 1,000 yuan/household; for livestock sheds and production materials: 1,000 yuan/household; for housing construction: 30 m^2/household × 350 yuan/m^2 = 10,500 yuan; for livelihood assistance and production commencement expenses: 4,500 yuan/person × 4 persons/household = 18,000 yuan. Thus, a total of 30,500 yuan is provided in grants-in-aid per household.

VIII. Analysis of the effects of ecological migration

Ecological effects

The implementation of ecological migration programs produces remarkable ecological effects. For example:

(1) After migrants have moved from their land, large areas of grassland and arable land can be used for "forest and grassland restoration." That dramatically reduces the ecological damage caused by humans and livestock in these areas. Allowing the land to "recuperate" contributes to recovery of the grassland vegetation.
(2) Through the protection, management, and rehabilitation work of stationed staff, the trend of desertification and land degradation can be effectively restrained. In addition, the recovery of vegetation can be accelerated, and the occurrence of natural disasters such as sandstorms and drought can be reduced. In short, these measures enable ecological conditions to be improved at the root level.

(3) The migration of herders from the "Helan Mountains National Nature Reserve," the "Suosuo Woodlands Reserve," and the "Poplar Woodlands Reserve" contributed to the recovery and growth of vegetation in these reserves. This recovery helps the forests to develop the ecological ability to heal themselves.

Farmers and herders relocated via the ecological migration program have implemented a series of small-, medium-, and large-scale comprehensive agricultural development projects in Luanjingtan, Yaobatan, the open sea of Chahar, and along the Ejene River, and have constructed artificial oases. As a result, the environment in these areas has been improved. The relocation of herders living under difficult conditions because of deteriorating grasslands enables the degraded grassland to recuperate; accordingly, the vegetation in these areas can be gradually recovered. In 1999, the Alasha League Government relocated all the 560 herder households who were engaged in grazing within the Helan Mountains National Nature Reserve. As a consequence, the vegetation in the area is clearly being recovered.

Economic and social effects

(1) Migration also contributes to relieving the poverty of farmers and herders. Ecological migration enables farmers and herders to significantly improve their incomes and greatly reduces the proportion of people living in poverty. Due to badly difficult living and livelihood conditions, the standard of living of the majority of farmers and herders prior to migration was below the poverty line. The government thus found it extremely difficult to conduct poverty relief projects. Merely giving away money and goods was insufficient to solve the problem at its root. Yet, as Alasha League is so vast and sparsely populated, there are areas of abundant water, land, and mineral resources where relatively large numbers of people can be accommodated. By implementing ecological migration and poverty relief migration to such areas as strategies for improving the lives of poor herders who suffer low standards of living as their subsistence is eroded by desertification and grassland degeneration, the herders are given an opportunity to escape their poor environment. In addition, government support and the migrants' own efforts to help themselves have achieved poverty relief in a short period of time, with the problem of poverty solved at its root. As a

result of the intensive development and poverty relief achieved through migration, the number of people living in poverty in livestock-raising areas fell from 23,000 persons in 1993 to 10,100 persons by the end of 1999. Over this time, the proportion of people living in poverty dropped from 49.7 percent to 15 percent. Net income per farmer or herder rose from 991 yuan in 1994 to 2,284 yuan in 1999. In the six years from 1994 to 1999, 9,300 ha of irrigation land was developed, 1,687 motors and electric wells were built, 1,300 ha of low-yield farmland was improved, 2,000 silos were built, over 200 livestock for floating poverty relief were organized, drinking water for an additional 38,400 persons and 83,600 livestock was secured, national highways were opened in six sums, and electricity was introduced in 27 sums.

(2) Migration also contributes to changing production methods in farming and livestock raising and the rational adjustment of industrial structures in livestock raising areas. Large areas of Alasha League were used for grazing, while farming was conducted in some parts of the league. The league could well have been characterized as a relatively closed peasant economy. Because of backward production methods and ecological destruction, it was impossible for the inhabitants to increase their income. The areas of higher population density that developed with migration made it easier to initiate construction in these areas and for large-scale management practices to be implemented. This construction and management contributed to the rational utilization of land resources and the restructuring of farming and livestock economies. In addition, "high-yield, high-quality, and high-efficiency" farming and livestock raising were developed in several ways. For example, adapting scientific farming and livestock raising management methods such as feedlot raising based on water-efficient irrigation and feed crop cultivation, as well as intensive feedlot raising; applying state-of-the-art science and technology; and making use of modern technology for communication, transportation, and distribution, etc. In this way, productive cycles, in which farming and livestock raising promote each other, have been established; major reforms were made to the industrial structure; economic development was accelerated; and the evolution of farming and livestock production methods was promoted. In 1999, the output of farming and livestock production amounted to 523.227 million yuan, at an annual increase of 13.2 percent. Of this, farming (crop cultivation) production represented 248.419 million

yuan, an increase of 19.25 percent. Following large-scale economic restructuring in livestock-raising areas, the number of people engaged in livestock raising fell from 14,096 households in the 1980s to 9,039 households in 1999. The proportion of farming in primary industry production increased from 38.15 percent in 1993 to 47.78 percent in 1999. Feedlot raising in farming areas increased from 30,000 head of livestock in the 1980s to 230,000 heads in 1999. The internal structure of primary industry was also improved; the process of industrialization was accelerated, and a series of new enterprises experienced rapid growth, including the production, processing, and sale of cashmere, the production and sale of watermelon, and the production and processing of grass and seed.

(3) Ecological migration contributes to the economic unification of cities and rural areas, achieving a rational reallocation of labor resources and accelerating the construction of small cities. The herders that were previously engaged in natural grazing, living scattered throughout the vast deserts, Gobi, and natural forest areas, were forced to rely on nature for their livelihood. They were completely separated from society and thus lacking in education. Following migration, they found themselves living in cities and development districts of relatively high-population densities; this relocation accelerated the construction of small cities. Migration was assisted in the implementation of cultural, educational, and scientific and technological projects. Thus, the economic unification of cities and rural areas was achieved, allowing the migrants to experience and enjoy modern and civilized living. Small cities in livestock-raising areas began to grow quickly as a result of migration. A number of small cities, each equipped with a series of functions, have taken shape, including Jargalsaihan (Jiaergalesaihan) Town (Luanjingtan), Baruunbel (Barunbieli), Usutu, Unduraltu (Wenduerletu), and Yabrai (Yabulai). The urban population grew from 49,968 persons in 1980 to 115,600 persons in 1999. The proportion of people living in urban areas over the entire league reached 66.36 percent. Along with the development of small cities, a market economy also developed. Each small city has given rise to markets for farm produce and trading markets of various sizes. In addition, market information is now transmitted quickly between different areas. Township companies and town companies have also emerged. The gross product (per capita) of the league reached 10,966 yuan in 1999, with an average

annual increase in recent years of 56.29 percent. The town and township companies are having an active and important impact on the income of farmers and herders.

(4) Ecological migration is also contributing to the sustainable development of the social economy, social equity, and social stability in minority ethnic districts.

The voluntariness of migration under the "ecological migration" policy: from case studies of herders in Ordos City, Inner Mongolia Autonomous Region

SHUNJI ONIKI and B. GENSUO

Introduction

The desertification of grasslands around Ordos (Eerduosi) in Inner Mongolia (Neimenggu) Autonomous Region of China has been progressing for many years, due to agricultural expansion, increase in firewood collection, and overgrazing that has accompanied population growth. (Yoshikawa, 1998: 34). Shi et al. (1998) argue that in order to improve the standard of living of herding households in areas of severe desertification, it is important to rehabilitate degraded grazing lands without reducing household income; to achieve this, they argue that, it is necessary to increase the use of feedlot-style livestock raising. Based on such concepts, an "ecological migration" policy, aimed at relocating herders from grassland areas to houses with attached feedlots in areas adjoining cities, is being introduced in various areas of Inner Mongolia.

Although people are forced to relocate in many cases of "ecological migration," there is considerable variation in these cases throughout the country. As yet, there is insufficient evidence to prove that forced migration is the most effective method of applying this policy. In Wushen Banner, in Ordos, "ecological migration" is being carried out in a relatively flexible way; for example, some of the inhabitants are allowed to exercise free choice regarding migration. In this chapter, we consider a case study from the area and attempt to elucidate the most effective method of implementing "ecological migration" from an economic perspective.

Cost effectiveness of migration policy

In Wushen, in Ordos, both forced and voluntary migration are taking place. In the case of forced migration, all herding households in a designated area are relocated to migrant villages and provided with a certain amount of compensation, in the form of houses, farmland, and livestock, to make up for losses caused by migration. Voluntary migration begins when the government announces details of the compensation to be provided to migrants; after seeing this, households wishing to migrate submit an application and relocate.

Apart from the difference in application procedure, forced migration and voluntary migration also differ greatly in their economic effect. Normally, there are significant differences between individual herding households in terms of their management skills, family composition and the environmental conditions of their grazing lands. For this reason, some households bear a great deal of loss for the migration, while others may lose nothing. For example, the burden of migration is relatively light for people who are easily able to find secondary work after moving to the outskirts of a town,[1] whereas for someone who speaks only Mongolian, moving to a migrant village in a town with a predominantly Chinese culture represents a relatively heavy burden. In this context, "burden" does not mean only monetary costs, but also psychological and spiritual hardships. If a household with a relatively light burden of migration moves instead of a household with a heavy burden, the effect on the environment will be the same, but the total burden – that is, the total social cost for the area – is minimized. In this sense, voluntary migration is a cost effective policy than forced migration, because it has a lower social cost.[2]

Since it is generally not possible to know to the extent of loss of utility (income or other social welfare) of a herding household, it is difficult for government officials to calculate precisely how much compensation should

---

1 This is a concept of environmental economics. In this case, the cost of giving up cultivation of one existing unit of livestock (the additional cost of cutting production by one unit) is known as the marginal abatement cost. Under a subsidy policy in which livestock producers participate voluntarily, producers with lower marginal abatement costs tend to abate production more readily. Compared to compulsory livestock reduction, this results in lower costs for the whole of society.
2 Here, "cost effectiveness" means that the same effect can be achieved at a lower total cost.

be provided in the case of forced migration.[3] However, in voluntary migration the problem of insufficient compensation does not arise, because each individual household has a chance to compare the advantages and disadvantages of migration and then makes a decision on whether to move. If the members of a household find, after migrating, that the benefits are less than anticipated, they can return to their original land. In this way, voluntary migration has an added advantage in that it enables accurate measurement of the effects of migration policy.

However, compulsory migration also has the advantage that the policy can be implemented promptly and with certainty in areas of severe land degradation that require immediate relocation of all inhabitants. If management styles and preferences are homogeneous among headers, such that the effect of migration was the same for all households, and if the government knew precisely the amount of compensation required by the herders, then there would be no difference between forced and voluntary migration in terms of cost effectiveness. However, from the point of view of policy implementation procedure, certainty of results, and speed, forced migration seems to be the better option.

Accordingly, in order to decide which type of migration policy is implemented, it is necessary to know the degree of individual difference among the inhabitants with regard to migration, and also whether the government can make an accurate estimate for the compensation required

An outline of "ecological migration" in Wushen

General geographical situation

The survey area for this research was the northern part of Wushen Banner, located in the southeastern area of the Muu us (Maowusu) Desert. It is a semi-arid region, with annual precipitation of approximately 350mm. The

[3] Here, "utility" means not only financial benefit, but various other values that contribute to individual welfare. The effects on individuals who move to the "ecological migration villages" cannot be evaluated solely in financial terms – they must also include satisfaction or dissatisfaction with leading an urban life. For details on the evaluation of such effects, refer to, for example, Varian, 1999.

topography is desert grassland with a mixture of sand dunes and thickly growing plant populations (Yang 2001: 15–19). The groundwater level is high and the disappearance of water sources is not evident. Because the ground is composed of fragile layers of sandstone, which easily turn into sand, and because of the destruction of vegetation due to human activity, an area of the desert has been extended. As the population rose quickly, many trees were cut down for fuel and other uses, and the number of livestock grew rapidly within limited areas. In addition, areas of grassland were repeatedly cleared and then abandoned. In this way, desertification has progressed (Uchimouko Sabaku Kaihatsu Kenkyukai Hen, 1989).

Between 1949 and 1990, the population of Wushen Banner became more than three times and the number of small livestock became about four times (Wushen-Qi Zhi Bianji Weiyuanhui bian, 2001). By the mid-1970s, the area of the Muu us Desert had become more than three times (Uchimouko Sabaku Kaihatsu Kenkyukai Hen, 1989).

The gross product per agricultural worker in Wushen Banner in 2002 (7,736 yuan) was slightly higher than the average for the Inner Mongolia Autonomous Region; as a whole, the banner is not necessarily poor on average. However, the proportion of people living in poverty as of 1986 was 17 percent, (Wushen-Qi Zhi Bianji Weiyuanhui bian, 2001) so it seems there is a considerable number of poor farmers and herders.[4] In Wushen Banner there are large-scale industrial plants whose operations are based around the natural gas produced in the area, which drives up industrial output (Inner Mongolia Autonomous Region Bureau of Statistics, 2003). However, as these are capital-intensive plants, the proportion of the banner population employed in this area is low; the majority of people are still employed in primary industry.[5] Thus, the farming households in this area are not blessed with many off-farm employment opportunities.

---

4 Gross product per person is 12,911 yuan, which is higher than average for the Inner Mongolia Autonomous Region (7,233 yuan). "Poor households" are defined as those having a net income of less than 120 yuan per person per month in farming areas and less than 150 yuan in herding areas.
5 The proportion of people engaged in primary industry is 71 percent and that engaged in secondary industry is 12 percent (Inner Mongolia Autonomous Region Bureau of Statistics, 2003).

The "ecological migration" project

The construction of "ecological migrant villages" in Wushen Banner began in 2002; it is expected that a total of 1,000 houses for migrants will have been constructed in various areas of the banner by the end of 2004. This housing capacity corresponds to about 3.5 percent of the rural population of the banner. As of August 2004, approximately 400 of these houses were already occupied.

Relocation takes place for some migrants under the "ecological migration program", while others are relocated as part of the poverty alleviation program *(fupin)*. Under the "ecological migration" program, migrants are entitled to receive 4,500 yuan per household at the time of relocation; in the poverty alleviation program, the entitlement is 5,000 yuan. In reality, few migrant households actually receive the subsidy incash, perhaps because the funds are used for some other purpose. In fact, the migrants are required to pay a lump sum of 3,000 to 4,000 yuan at the time of relocation.

The migrant village in which we conducted our survey is located about 100 km north of the center of the banner. As of August 2004, 104 houses had been completed and 30 households had already moved into the village.

The buildings in the village are designed to accommodate two households each. There are houses for raising sheep and pigs on one side of the main road and those for cattle on the other side. On average, each household has approximately 80 m^2 of a feedlot with a barn for raising livestock, a residence of approximately 30 m^2, and arable land of approximately 0.67 ha (10 *mu*). According to the farmers themselves, it is possible to keep up to 50 sheep, 7 cattle, or 30 pigs with these facilities. At present, as households are typically raising only 20 to 30 sheep, 1 to 2 cattle and up to 10 pigs, there is still room for them to expand their livestock holdings.

As a matter of policy, the banner government organizes two visits a year to the village to provide livestock management guidance and veterinary services to the migrants. However, information is not reliably conveyed, and because many of the ethnic Mongolians cannot speak Chinese, a considerable number of households are receiving insufficient guidance[6].

Decisions on the location of migrant villages and the households to be relocated are taken by the banner government. Some of the more important factors examined in making this decision are convenience of transportation,

---

6   The language barrier is a major obstacle in the education of Mongolian children. One reason that some herders do not wish to migrate is to allow their children to attend schools that teach in Mongolian.

conditions for crop cultivation, provision of water and electricity supplies and telephone lines, and availability of market information. In instituting the process of migration, the area from which people are to be relocated is selected, and then a call is made for volunteer migrants. Depending on how this process is carried out, it may be perceived as compulsory migration or voluntary migration. In most cases, after migration, grazing bans are imposed on areas from which the inhabitants have been relocated. Migrants are permitted to cut grass in these areas, but grazing is prohibited. In addition, it is required that trees be planted in the vacated area.

Income structure of migrants

The intentions of migration and the incomes of migrants

In order to investigate the effect of the current migration policy on household income, we conducted a survey of the livestock management of 21 households in the migrant village. Surveys were conducted in March, July, and August 2004. Household incomes before and after migration are shown in Table 8-1. Here, "household income" refers to the total household income excluding subsidies and gifts, *i.e.*, the total agricultural income plus income from off-farm work. Agricultural income is calculated by subtracting operating expenses from gross incomes[7]. The migrants in this village frequently return to their old dwelling areas, and some stay there temporarily. As their activities in the village and in the old areas are combined as a single operation, their household incomes include earnings from both places. The disposable income shown in the tables is calculated by subtracting heating and lighting expenses – the cost of items such as electricity and coal – from household income.

7 Interest on financial assets and remittances from other families are not included in income. Interest on loans, family labor costs, land rentals, and taxes are not included in the calculation of expenses. Subjects of the survey included people who had just moved to the village and people who had bought or sold large numbers of livestock at the time of migration. Thus, for the purpose of income calculations, we used the average values obtained in another survey on 117 herder household in Uushin Banner that we conducted in 2004.

The average agricultural income per household after migration was 9,457 yuan; this is added to income from off-farm work (2,820 yuan) for a total average household income of 12,277 yuan. Comparing average incomes before and after migration, we can see that agricultural income fell by 3,312 yuan (25.9 percent of household income before migration), while off-farm income rose by 2,820 yuan. The net result was that household income fell by 492 yuan (a drop of 3.9 percent). Because their consumption of fuels such as coal increased when people moved to the migrant village, disposable income, as calculated by subtracting heating and lighting expenses, fell by 1,901 yuan (a drop of 14.9 percent).

Table 8-1. Income breakdown of migrant village residents (yuan)

	Before migration	After migration
Agricultural income	12,769	9,457
Off-farm income	0	2,820
Household income	12,769	12,277
Disposable income	12,724	10,823
Change in agricultural income		-3,312
Change in disposable income		-1,901

Table 8-2. Types of migration and intentions of migrant village residents

	Voluntary migration	Compulsory migration	Want to return	Do not want to return	Total
Change in agricultural income (yuan)	-3,140	-4,347	-4,259	-2,730	-3,312
Off-farm income (yuan)	3,235	333	1,500	3,633	2,820
Change in disposable income (yuan)	-1,512	-4,240	-2,994	-1,229	-1,901
Age of household head	44	42	41	45	43
No. of household members	3.53	3.67	3.25	3.75	3.55
No. of livestock before migration	70.9	78.3	64.8	76.4	72.0
Proportion of desert in the total land area (%)	46.6%	53.0%	51.0%	45.3%	49.9%
Grazing density (no. of animals/*mu*)	0.30	0.26	0.42	0.22	0.31

Comparing the incomes of the migrants with the average incomes of livestock herders and farmers in Inner Mongolia Autonomous Region, we find that the household income of migrants after relocation is slightly lower than that of herders, but higher than that of crop farmers. For agricultural households in Inner Mongolia in 2003, the average agricultural income per head was 1,548 yuan and the average off-farm income was 340 yuan; for herding households, the average agricultural income per head was 2,715 yuan and the average off-farm income was 179 yuan.[8] For inhabitants of migrant villages, agricultural income per head was 2,664 yuan and off-farm income was 794 yuan. From this we can conclude that the household incomes of migrants are not significantly lower than average for agricultural households in Inner Mongolia, but the proportion of income earned from off-farm sources is much higher than average.

Next, we divide the survey households into those that underwent compulsory and voluntary migration, and compare the average incomes of the two groups. Then, we find that the drop in income was smaller in the households that undertook voluntary migration (Table 8-2). This can be explained by the fact that voluntary migrants often have higher off-farm incomes. Those who volunteer for migration tend to be confident of finding off-farm employment, and if they find the work opportunities to be less than expected, they may return to their previous homes. From our interview with migrants, we found out that the most common reason for wanting to return is a failure to obtain the expected level of work. In this survey we did not observe many differences based on age, number of livestock or grazing pressure; the most important factor in the decision to migrate was whether sufficient off-farm income could be earned.

Inputs of production

The migration of herders to "ecological migrant villages" has resulted in a technological shift from extensive livestock raising, based on natural grazing, to capital-intensive livestock production in feedlots. This technological shift alters the demand for labor, that is, there are fewer hours of neces-

---

8 Inner Mongolia Autonomous Region Bureau of Statistics, 2003: 252–253. "Household net income" and "labor remuneration." Here, "household income" is equivalent to "agriculture net income" in the statistics yearbook, and nominal prices are listed.

sary labor, but the supply of labor, which corresponds essentially to the number of family members, is difficult to change in a short period of time. Thus, a mismatch occurs at least in the short run between labor supply and demand, resulting in surplus labor. Then, unless the surplus labor is productively used, household income will clearly drop.

Most of the households that we surveyed reported that they needed fewer hours of labor for raising livestock after migration (Table 8-3). As well as a reduction in the number of livestock, the drop in labor demand is largely to do with the fact that the labor-intensive work of grazing livestock is no longer necessary. The surplus labor resulting from migration is often directed to off-farm work. Approximately 80 percent of surveyed households contained people who were utilizing their surplus time by doing off-farm work or looking for such work (Table 8-4). This means that the income of a household may increase or decrease after migration, depending largely on whether it is able to effectively utilize its surplus labor.

Table 8-3. Change in agricultural labor time after migration

	Proportion
Reduced	55.6%
No change	38.9%
Increased	5.6%

Table 8-4. Utilization of surplus time

	Proportion
Off-farm work	53.3%
Do not work	20.0%
Looking for work	26.7%

Table 8-5. Debt situation

Proportion in debt (%)	57.1%
Average debt (yuan)	7,000.1
Debt to (%)	
Relatives	45.5%
Banks	27.3%
Cooperatives	20.0%
Governments (projects)	9.1%

The shift to intensive livestock production generally demands higher value of livestock, but at present the migrants possess insufficient stocks of animals. Although raising a greater number of livestock would make use of surplus labor and lead to increased income, people do not have enough capital to purchase livestock. The households we surveyed procured funds mainly by borrowing from relatives, but such loans are not necessarily sufficient for expanding livestock operations: the average loan amount is 7,000 yuan (Table 8-5), but this amounts to only half the price of one dairy cow. Few financial institutions offer unsecured loans. The ability to procure finance is a major issue in efficient livestock management by migrants.

Each household in the migrant village is provided with agricultural land for growing feed crops. This quantity of land, per household, corresponds to about one fifth of the average area of land (per person) for agricultural workers in Inner Mongolia.[9] Based on the average yield of feed corn in the local area of the migrant village, this amount of land produces enough feed to raise 30 breed-improved sheep in a feedlot.[10] If efficiently utilized, this is the minimum size required to produce feed crops needed for the migrants' livestock.

Case studies of livestock cultivation

Up to now, the discussion has dealt with average values, but here, in order to show more concretely the actual conditions for "ecological migrants," we present some case studies of households who undertook voluntary migration.

Mr. "A" (31, Mongolian), a pig farmer, lives with his wife in the migrant village. In the area where he used to live, approximately half of the herders migrated under the ecological migration program. The area has

---

9 This assumes an average labor of two people. The average value for the Inner Mongolia Autonomous Region is taken from the Inner Mongolia Autonomous Region Bureau of Statistics,, 2003: 285.

10 This calculation is based on the assumption that the improved breed of sheep in this area, *Xiaoweihanyang*, produces three offspring each year. The average feed corn yield in 2003 was 6,740 kg per hectare, the feed consumption of one mother sheep and three baby sheep is 1,611 kg per year (Ordos-Shi Xumuyeju, 2002), and the production ratio of corn to corn foliage is 1.67 (Liu et al., 2003).

been designated a "grazing ban" area, but his parents and a younger brother remained and are raising sheep in a feedlot. Upon migration, Mr. A sold 16 sheep, gaining 3,200 yuan of funds. He also borrowed 8,000 yuan from an acquaintance to use for some initial investment in the migrant village (Photo 8-1).

Photo 8-1. Inside a barn renovated for pig farming

Mr. A is currently raising eight pigs. He buys piglets for 150 yuan per head, and after six months he sells them for 700 yuan. In one year he feeds his pigs three tons of feed corn, costing 1.2 yuan per kg, and 1,500 yuan worth of formula feed. After subtracting the cost of his material input, his profit from pig farming in one year amounts to 3,700 yuan.

Mr. A also does some secondary work occasionally, such as casual work at a nearby factory, agricultural work for other farmers and construction work. When he is occupied with secondary work, his wife looks after the livestock. He earns 25 to 30 yuan per day from his secondary jobs, which amounts to about 6,400 yuan per year – more than Mr. A's agricultural income.

Mr. A plans to keep more pigs some time in the future. He feels that with some extra money, his current labor resources and facilities would permit him to raise up to 20 pigs. If he achieved this, he could increase his agricultural income to 9,250 yuan per year.

The case of Mr. A is an example of part of a large family, which had previously lived quite poorly, increasing its income after migration by a combination of pig farming and secondary work. However, it is important to consider the risk Mr. A is taking in his business. Many crop farmers in China engage in pig farming as a sideline, using feed that they grow themselves. In contrast, full-time pig farming, as pursued by Mr. A, involves considerable cash expense for the purchase of feed and other items; thus, the economic risk is high (Yao & Uno, 1998). In order to create a more secure livestock operation, it is necessary to increase the proportion of self-supplied feed used. This may be achieved by utilizing the surplus labor of farming households to assist in the cultivation of feed crops, thereby reducing the cost of purchasing feed. Compared to unstable temporary employment, this would result in a more secure livelihood.

Mr. "B" (71, Mongolian) is a herder who migrated voluntarily to the village. Together with his wife, he is engaged in dairy farming. Due to his advanced age, he is not doing any secondary work. He is currently raising one dairy cow (Photo 8-2); if possible, he would like to raise one more cow.

His second-eldest son has remained with his wife in the area where Mr. B lived previously, raising 184 sheep, 9 cattle, 1 mule, and 1 pig (Photo 8-2). Many of the migrants who relocated from the same area at the same time as Mr. B have now returned there after failing to find secondary work. However, because Mr. B has a large family, even if he remains in the migrant village his children can continue to raise livestock where the family used to live, as he conciders that the area has not been designated a "grazing ban" area. Although Mr. B was given 0.67 ha of land in the migrant village, he has planted only 0.14 ha of feed corn. He claims the soil is sandy and poor. He also grows feed grass in the courtyard of his house, which sits on a plot of approximately 45 m^2. After migration, Mr. B's agricultural income, including that from his original land, dropped by 3,971 yuan, but under the Conservation Set-Aside Policy ("Tui geng huan lin"), he receives about 4,000 yuan as a subsidy, which compensates for his loss of income.

The number of livestock kept in the area where Mr. B originally lived fell by 16 percent (in terms of equivalent number of sheep) after migration. The total grassland area, including shrub growth and afforested areas, is only 60 ha, so even after this drop in livestock numbers, there are still too many animals for the available grassland. Assuming that 0.80 ha of grassland is needed for each sheep,[11] about three times more grassland would be needed in order to achieve an appropriate grazing pressure. Accordingly, unless the area of grassland available to each household is increased, grazing pressure will not significantly change. In order to reduce grazing pressure on grassland areas, another step is needed, such as rental of land of departed migrant households to remaining households.

Photo 8-2. Animal barn (behind) and house (front) of ecological migrants raising dairy cattle

---

11 This is the average value of the number of livestock that can be raised per unit area, calculated using the valid answers to a household livelihood survey on livestock raising.

The income of migrants such as Mr. B, who do not engage in secondary work and have few livestock in the migrant village, cannot be expected to increase. However, the migration of large families with surplus labor to villages allows some reduction in grazing pressure in the vacated grasslands without significant changes in household income.

Since Mr. A and Mr. B either have a large family or the skill to undertake a new kind of livelihood, they can both take advantage of the opportunity presented by migration. If migration is based on free will of herders, it is likely that many people like Mr. A and Mr. B, who can adapt well to relocation, will opt to migrate.

Implementation of an appropriate "ecological migration" project

"Ecological migration" creates pressure on migrating herders to greatly modify their economic activities. Herders who move to the outskirts of cities take up intensive livestock raising – with which they have had no previous experience – and, in some cases, take secondary jobs, doing the same kind of work as city laborers. While some herders can adapt to this type of urban life, others cannot. Accordingly, the policy is appropriate for herders who can easily shift to a new lifestyle, while those who cannot do so must bear a huge, unseen cost. The provision of sufficient compensation by the government enough to satisfy all herders for whom migration presents a heavy burden would require an extremely large financial expenditure. However, if herders are forcibly relocated without receiving satisfactory compensation, the invisible cost will be only shifted from the government to the herders; the social cost does not simply disappear. The key to conducting "ecological migration" in a cost effective way is to promote a form of migration based on free will of herders. If insufficient numbers of people volunteer for migration, the government will be obliged to improve the compensation offer. Even then, the social cost would be less than that of forced migration. In this survey, we found that under a system in which migration is optional for herders, it is those households which are capable of increasing their incomes that tend to migrate;[12] hence, the social cost of migration is lower for voluntary migration than for compulsory migration.

Furthermore, in order to decrease overall grazing pressure on grasslands after voluntary migration, the per-household grazing area of the remaining herders should be expanded by promoting the leasing of land between herding households. At present, areas of land in which grazing has been discontinued form a mosaic with areas where grazing continues, with the effect that the limited overall amount of grassland is not being utilized efficiently. If herders remaining in the grasslands were to use the entire available area, a reduced number of livestock could graze evenly over a wide area, resulting in a more appropriate grazing pressure; in this way, they could prosper economically while reducing grazing pressure on the grasslands. Meanwhile, if people who migrate do so willingly, it is because they have something to gain by migration. In other words, there are benefits for both the people who migrate and the people who do not migrate.

In order to implement voluntary migration for a local government with limited finances, some innovative initiatives are necessary to help migrants increase their incomes. As mentioned earlier, the inevitable problem associated with "ecological migration" is that the switch to intensive livestock raising results in surplus labor. The labor necessary to keep grazing animals is not needed after migration; instead, there is an increase in operating expenses, such as the cost of purchasing feed and other items, so that overall income falls. To prevent a reduction in household income, surplus labor must be absorbed by cultivating feed crops or increasing livestock numbers, or in off-farm employment. Under current conditions, off-farm employment is an important factor in maintaining household income, but in the outskirts of small towns, opportunities for off-farm employment are limited. Accordingly, it is important to direct surplus labor to activities related to livestock raising, such as cultivating feed crops and expanding the scale of operations.

For the future, in order to increase the incomes of people in migrant villages, it is important to overcome the problems of insufficient capital and land and insufficient technical knowledge and information. First, as migrants tend to possess relatively few livestock before migration and lack

---

12 In some cases, grazing bans were implemented in the areas from which people had migrated. For this reason, we could not explore changes in income without including the influence of grazing bans and the impact of migration policy.

the capital to cover the initial investment needed to establish a feedlot operation, financial support for migrants who wish to purchase new livestock would be desirable. Second, migrants should be provided with good-quality arable land of sufficient size to enable them to cultivate enough feed for raising animals in a feedlot. Third, a system for providing detailed technical guidance should be set up to help herders adapt to new production methods and continue efficient production; we were told that some herders in the migrant village had been provided with breed-improved sheep but ended up selling them because they did not know how to raise the animals properly. We hope for better opportunities for communication between the government and migrants, and the establishment of an improved support system based on the real needs of the migrants.

References

Inner Mongolia Autonomous Region Bureau of Statistics, 2003, *Inner Mongolia Statistical Yearbook 2003*, China Statistics Press.

Liu Jianning, He Dongchang, Wang Yunqi, Liang Quanzhong, Mao Yangyi and Zhang Lijun, 2003, *Beifang Ganhan Diqu Mucao Zaipei yu Liyong*, Jindun Chubanshe. {Chinese} (Liu Jianning, He Dongchang, Wang Yunqi, Liang Quanzhong, Mao Yangyi and Zhang Lijun, 2003, *Cultivation and Utilization of Pastures in the North Dry Region*, Jindun Publishing House.)

Ordos-Shi Xumuyeju, 2002, *Biaozhunhua Shesiyangchu Jishu Caozuo guicheng*, Erduosi-shi. {Chinese} (Ordos-shi Bureau of Animal Husbands, 2002, *Regulations on Manipulation of Stock Farming, which Normally Conducts Livestock Farming in Barns*, Ordos-City.)

Shi Minjun, Tanaka Yosuke and Zhao Halin, 1998, Nobokuchiiki ni okeru tochiriyo no tenkai to sabakuka mondai – Chugoku Horuchin-Sachi no Jirei, *Tsukuba Daigaku Nourin Shakaikeizai Kenkyu*, 15, pp. 1–26. {Japanese} (Shi Minjun, Tanaka Yosuke and Zhao Halin, 1998, Land use and desertification in interlacing agropastoral areas of North China – A case study on Korgin Sandy Land, *Memoirs of Institute of Agriculture and Forestry, the University of Tsukuba. Rural Economic and Sociology*, 15, pp. 1–26.)

Uchimouko Sabaku Kaihatsu Kenkyukai hen, 1989, Chugoku no kansoiki ni okeru sabakuka no kikokaimei to dotaikaiseki – Maowusu-sabaku no sabaku ryokuka to nogyo kaihatsu ni kansuru kisoteki kenkyu, *Toyota Zaidan Josei Kenkyu Hokokusho 012*, p. 15. {Japanese} (Research Group of Development of Desert in Inner Mongolia,1989, Analysis of mechanism and movement of desertification in the arid land areas in China – basic studies on desert greening and agricultural development in the Mu Us Shamo Desert of the Inner Mongolian Autonomous Region, *The Toyota Foundation Grant Research Report 012*, p. 15.)

Ueda Kazuhiro, Oka Toshihiro and Nizawa Hidenori, 1997, *Kankyoseisaku no Keizaigaku – Riron to Genjitsu*, Nippon Hyoron Sha. {Japanese}

(Ueda Kazuhiro, Oka Toshihiro, Nizawa Hidenori, 1997, *Economics of Environmental Policy – Theory and Reality*, Nippon Hyoron Sha.)

Varian Hal R. 1999, *Intermediate Microeconomics: A Modern Approach*, 5th edition, W.W. Norton & Company.

Wushen-Qi Zhi Bianji Weiyuanhui bian, 2001, *Wushen -Qi Zhi*, Neimenggu Renmin Chubanshe. {Chinese}

(Editorial Board of Annal of Wushen, 1998, *Annals of Wushen Banner*, Inner Mongolia People's Press.)

Yang Haiying, 2001, *Sogen to Uma to Mongorujin*, Nippon Hoso Kyokai. {7Japanese}

(Yang Haiying, 2001, *Grassland/Horses and the Mongolian*, Japan Broadcasting Corporation.)

Yao Fengtong, Uno Tadayoshi, 1998, Chugoku no nokafukugyoteki yotonkeiei no jittai to mondai, *Nogyo Keizai Kenkyu*, 70(1), pp. 36–46. {Japanese}

(Yao Fengtong, Uno Tadayoshi, 1998, Conditions and problems of pig raising industry as a side line business in China: Case study in Inner Mongolia, *Journal of Rural Economics*, 70(1), pp. 36–46.)

Yoshikawa Ken, 1998, *Sabakuka Boshi he no Chosen – Midori no Saisei ni Kakeru Yume*, Chuo Koronsha. {Japanese}

(Yoshikawa Ken, 1998, *The Challenge of Prevention from Desertification – The Desire to Recover Greenery*, Chuo Koron Sha.)

III *Questioning cultural aspects: What kind of transformation does "ecological migration" effect?*

Cultural acceptance of inhabitants in "ecological migration" from case studies in Xianghuang Banner, Shilingol League, Inner Mongolia Autonomous Region

ALTA

Introduction

It is estimated that there are approximately seven million people now living in poverty in rural China, due to difficulties in making a living from deteriorating grasslands. A sizable number of these people are living in Inner Mongolia and other parts of western China. To address this concern, the national government has increased investment in these areas and is implementing an "ecological migration" plan that aims to improve the standard of living of the poor by relocating them from degraded ecological environments.

By the end of the 20th century, 2.2 million people, out of a total of more than 3.0 million living in poverty in the Inner Mongolia Autonomous Region, had managed to escape poverty. However, 1.5 million people fell below the poverty line again due to a drought that began in 2000, and in 2001 a further 990,000 people joined the ranks of the poor, bringing to 3.94 million the total number of people living in poverty in agricultural and livestock-raising areas. Of these, 1.5 million were officially designated as "especially poor" (Wuligeng, 2003: 12–14). At this point, the government of the Autonomous Region invested several hundred million yuan in a plan to effect the migration of 650,000 people over a period of six years. Of the 82 administrative units (corresponding to county level) in the Autonomous Region, 72 are involved in this plan (Fang, 2003: 4–6). According to the relevant sections of the Inner Mongolian government, migration projects are being implemented by various levels of government in the Autonomous Region. In 2002 and 2003, 14,000 and 18,000 people, respectively, were relocated, and a further 20,000 people are expected to migrate in 2004.

In the "pure" livestock raising areas of Inner Mongolia, traditional nomadic herding was practiced in some areas until the early 1990s. However, due to government changes to the grassland utilization system, nomadic herders gradually gave up their traditional practices and took up "settlement livestock raising." With the current boom in ecological migration, the lives of herders are undergoing further changes. The traditional style of grassland grazing is being replaced with year-round feedlot raising, or a method that combines feedlot raising with natural grazing; some herders are relocating elsewhere to engage in dairy farming or work in non-livestock-related jobs.

It is widely held that the basic cause of degradation of grassland environments in China has been the growing population of herders and the resultant overgrazing due to increased livestock numbers in traditional herding (e.g. Hou, 2002: 63–69). Thus, to reduce population pressure and to divert herders from traditional livestock raising, the government has systematically directed herders into other industries, such as dairy farming, feedlot-style livestock cultivation, and secondary and tertiary industries, and then banned grazing in the affected grasslands.

In May 2004, I conducted a preliminary survey in Xianghuang (Huveetushar) Banner, in the Inner Mongolia Autonomous Region, followed by an intensive survey of people relocated from the banner at an ecological migrant village in Hohhot (huhehaote) city, the capital of the Autonomous Region. In this paper, I describe the situation pertaining to ecological migration in the banner, and then present several case studies. While considering the adaptation of herders who were forced to migrate to the changes in their livelihoods, I examine the issue of maintenance of their traditional culture. Note that the administrative units of the Inner Mongolia Autonomous Region, in decreasing order, are: league (city), banner (county), sum (township), and gachaa (village).

Migration in Xianghuang Banner

Xianghuang Banner covers a total area of 4,960 km^2, and has jurisdiction over three townships and one town. Mongolians make up 64 percent of a total population of 27,300. It is a "pure" livestock-raising banner in which

Mongolians are the principal ethnic group and the main industry is grassland livestock grazing. According to the 10-year plan being implemented by the banner government, a total of 5,591 people will be relocated as "ecological migrants," and a total of 2.033 million *mu* (approx. 136,000 ha) of land will be set aside for discontinuation of grazing to restore the grasslands. This large-scale plan involves 20 percent of the total population and 27 percent of the total land area. There are four main methods of implementing the migration plan. One is to expand towns, relocate migrants to these towns, and develop dairy farming industries; the second is to build new migrant villages in township centers and areas in which water resources are lacking, and relocate the migrants to these; the third is to give the herders no alternative to giving up livestock herding and taking up employment in secondary or tertiary industries; and the fourth is to relocate migrants to different areas to engage in dairy farming or other industries.

Let us now consider the results of migration in this banner in 2002 and 2003. According to information provided by the banner government, a total of 200 herding households were relocated in 2002. Of these, 62 households are currently engaged in dairy farming on the outskirts of Hohhot city; 30 households migrated to the area near the railway station in Sholoon-huveetu-chagan Banner (on the Jining-Tongliao railway line which runs through the grasslands of central and eastern Inner Mongolia) to engage in railway-related work; 93 households were relocated to four newly constructed "ecological migration model parks" within the banner to engage in dairy farming and feedlot livestock raising; and 15 households relocated to the center of the banner, or outside the banner, to take up employment in secondary or tertiary industries. In addition, spring grazing bans were applied for two months in 2002 – from April 5 to June 5 – in 10 of the 60 gachaas in the banner. This measure affected 2,457 people from 749 herder households, a total of 93,000 livestock (in "sheep units"), and a total land area of 1.125 million *mu* (75,000 ha). The use of "sheep units" is a traditional Mongolian method for quantitative conversion between different types of livestock that uses a sheep as one livestock unit. Under this system, horses and camels are counted as six units, cattle five units, and goats 0.9 units. In 2003, year-round grazing bans were applied in 12 gachaas, affecting a total area of 666,000 *mu* (approx. 44,000 ha), and spring grazing bans were applied in 48 gachaas, affecting a total of 5.67 million *mu* (378,000 ha) of land. A total of 4,480 households and

527,000 livestock were affected by these bans, and overall, 98 percent of the banner has been subject to total or partial grazing bans. We can see that in these two years alone, the banner's migration plan has had an enormous effect on the lives of herders.

The situation of migrants

Migrant villages on the outskirts of Hohhot city

As mentioned, one of the destinations of migrants from Xianghuang banner is migrant villages on the outskirts of Hohhot city, the capital of the Autonomous Region. I will now consider in detail the conditions of migrants, with reference to case studies of migrants in one of these villages.

This migrant village is located on the outskirts of the city, a few kilometers from the city center. It occupies an area about the size of a sports field, and consists of one four-story apartment building, several single-story houses, and many fences. The site was previously used as a poultry farm, but after the farm went bankrupt, the bank operated a dairy farm here to try to recoup its losses. Since the site fulfills the five requirements set by the government for migrant settlements (roads, water supply, electricity, telephone and cable TV), and is also large enough to accommodate an entire settlement of herders, the government relocated migrants here to engage in dairy farming.

As mentioned above, 65 households of herders from Xianghuang Banner were relocated to Hohhot city in 2002 to engage in dairy farming. Of these, 17 households came to live in this migrant village. These 17 households represent half the total number of herders in their gachaa. There are also some households from other gachaas and other banners. The total number of households in the village is 22.

The migrants live either in the apartment building or in the single-story houses and are required to pay 1,100 yuan per year in rental charges for their residences and livestock sheds. For the 17 households from Xianghuang Banner, these charges are to be covered for three years by the Agricultural Development Bank of Hohhot as part of a poverty assistance program. The bank also supplies coal for fuel. A grant-in-aid of 500 yuan was provided by the government for the purchase of a straw grinder. The other households in the migrant village are required to cover all these expenses themselves.

In addition, the government provided each household with two Holstein dairy cows free of charge. After two years, some households had considerably increased the number of cows kept, while others had not. When these surveys were conducted, the number of dairy cows per household varied from a minimum of three or four up to a maximum of about a dozen. Annual income per cow was approximately 4,000 to 5,000 yuan. Milking is done twice a day at a nearby "milk station" which buys the milk and later sells it on to dairy processing companies. Peak milk production for dairy cows takes place over a three- to four-month period, during which a cow can produce 20 to 25 kg of milk per day. The annual average milk production is 12.5 kg per day. The milk station buys milk from the migrants for 0.43 yuan per kg, with payment on a monthly basis. As of May 2004, more than 100 dairy cows were being raised in this migrant village, producing in total about 2 tons of milk per day.

Feed for the cows is purchased by the herders themselves. Soon after arriving, the migrants set up a herders' cooperative, through which it was planned to purchase feed and sell livestock products, but in the end the cooperative did not achieve anything. The herders buy most of their feed as concentrated bulk purchases in the autumn, and then buy smaller amounts at other times as needed. In addition, the government provides a feed subsidy to each household which amounts to 5.5 kg per year per *mu* (15 *mu* is approx. 1 ha) of vacated land.

When these migrants left their gachaa they entered into a contract with the government, agreeing that they would not return to the grasslands for three years. However, due to the state of the grassland and the condition of the herders, the contract was extended for a further two years. The herders maintain that even if they return after five years they will not be permitted to conduct natural grazing as they did before. In the early 1990s, the grazing land in the gachaa was divided into lots of 300 *mu* (20 ha) per person and distributed among the herding households; after they left the gachaa, the government enclosed all their combined land with fencing.

Herding households who did not relocate to faraway areas to engage in dairy farming, but rather settled in migrant villages set up by the banner government within the banner, received 25,000 yuan each as a grant-in-aid from the government. In exchange, these herders are required to abide by grazing bans for two months of the year, during which time they must keep their livestock in sheds all day. Thus, herders who remained close to their homes are now raising their livestock under a "semi-feedlot" method.

The herders in the migrant village can be broadly categorized into two types: those relocated by the government, and "voluntary" migrants. I will present several case studies from this migrant village, obtained by interviews with some of the migrants, through which I hope to understand how the migrants' lives have changed since their relocation and to determine their plans for the future.

An example of a household whose livelihood has become more difficult – Case Study 1

Informant: Male in his 40s, living in a household of five, with his wife, a son, daughter-in-law, and grandchild

Originally I had more than 200 livestock. At the time of *otor* (a kind of migration), I entrusted 60 goats to the care of a friend and sold the rest, most of them for a very low price. I could only get 700 to 800 yuan for a cow with a calf. In 2001, when we left our homes, I got only 9,000 yuan for a total of 23 head of cattle.

I came here mostly because I followed the government's instructions, but also a little of my own free will. When we arrived, the government gave us two dairy cows, as promised. The government bought dairy cows in bulk from Shanxi province for a price of 16,000 to 17,000 yuan per cow. In this area I can buy a pregnant cow for 14,000 to 15,000 yuan per head. But I'm not experienced in raising dairy cows, so I can't tell the quality of a cow by inspecting its body. It's difficult to get a good-quality dairy cow. If you're unlucky you can find yourself with a cow that produces only a little milk. If you're really unlucky you can end up with a cow that can't even get pregnant. Then you have to sell it for beef. Anyway, I have fewer animals now, so it's hard to make ends meet.

If you ask me what I make of my life now, I have to say it's just OK. Some families have made a better life and some have it worse. It seems the families who had it better before tend to feel that it's not so good now, while the families who were worse off before tend to feel that things are a little better. That's because here you get a few livestock, and some cash every month.

I have to say I'm more or less used to my life here now. I can save a little from the money I make from the two cows, but it wouldn't be enough if I had a child going to school. If I could buy two or three more dairy cows, we could make a fairly comfortable living, but I don't have the money for that right now. At the moment I can only get milk from one cow.

It costs me at least 15 yuan a day to keep each of my cows. These cows are not as hardy as Mongolian cows, so if they get sick it costs a lot. Anyway, there are always unexpected costs. The cows are kept in a shed all day. They can't exercise, so they get diarrhea and other sicknesses quite easily. You can only get about 200 yuan by selling a male calf. It costs about 1,000 yuan to bring up a female calf for the first three months, for powdered milk and other essentials. Feed costs me about half a yuan per kg. One cow eats about 20 kg a day. I feed them three times a day, and I add cheap fodder, like corn straw, to the feed. If I could give them green stored feed their nutrition would be better. For now I rely on these two cows to make a living. Back where we lived before, I could just let the cows graze naturally in summer, without feeding them. The feed here is not nutritious.

After five years I plan to go back. I've heard that a milk station is going to be set up in the banner. Now I go back about once a year, in the autumn, to cut grass. But the cost of transport is high, so it's not convenient to go there and back. Nobody is living in our old house now. It's falling apart and I can't use it. The house is quite close to the town, so all the cow dung I'd stored up for fuel was stolen. I heard that the fenced-off grazing land is now managed by "ecology" people, but I saw that the grass was eaten up long ago by other livestock. The grazers only get fined if one of the ecology people runs into them. Most of the fencing at our house was stolen. It was Han Chinese who did all the stealing – maybe the price of scrap metal went up. The ecology people only manage the outside fencing; they don't look after the fencing of the houses inside. The grass inside the fence grows well, but when nobody is watching it gets eaten by the livestock of other herders in the area, so there's not much point in our doing *otor*. It would have been better if I had kept my 200 animals to graze on my land, not anyone else's land, and got someone to stay there and look after them. Going by what I heard, they're going to be stricter in controlling the grazing land in future.

I can make some kind of living now, but I worry that one day I won't be able to sell milk. Everyone's planning to go back to our village, but I worry just thinking about how I'll make ends meet there.

An example of a household whose livelihood has improved – Case Study 2

Informant: Male, in his 40s, with a family of six. Of his four children, three attend school back in their home village, and one is learning to be a cook in Hohhot city.

It's already more than two years since we came here. In this time I've gained a lot of experience. Looking after dairy cows is more troublesome than looking after the local cows. Quality is also a problem. If a cow is good you can get by somehow, but if not, it doesn't work out.

Right now I have three dairy cows and three calves. Two of the three cows were given to us by the government when we arrived; the other one I bought myself. It takes quite a lot of work to raise the calves until they produce milk. Each day I get about 50 kg of milk from my cows.

I got rid of most of the livestock we had before. I was quite satisfied with the price I got for them; in all, I got more than 10,000 yuan. But I still haven't rebuilt the capital I lost selling the livestock. I'm making about 2,000 to 3,000 yuan in a year, but I have a lot of expenses, like schooling for my children and medical costs.

Anyway, our situation now is slightly better than back in the village. We only have a few animals, so the work is easier. But the high cost of grass and feed is a problem. That's because we were grazing naturally before, whereas now we are using feedlots. What I'm hoping for now is that the milk station will offer a higher price for the milk they buy from us, because the price of feed and grass has increased.

I have no complaints about the dairy processing companies – I only have a problem with the milk station that buys milk from us. They sell the milk to the dairy processing companies for 1.35 yuan per kg, from which they make a profit of 0.2 yuan per kg. Raising animals in a feedlot is different from natural grazing, because we don't have to worry about snow and wind, it's hygienic, and management is detailed and efficient. The fact that so much milk is produced just in this village shows that we have a very advanced method.

The house where we lived before is ruined and so are all the things we had there. Everything was stolen by Han Chinese. If we did go back sometime, we would need to invest a lot of money. We had been given a total of 2,200 *mu* (approx. 147 ha), but after we came here all the grass on our land was eaten by the livestock of other households. That's because the banner doesn't manage the grasslands strictly enough.

If I can make a reasonable living here, I will stay on, but we're already reaching the limit of the amount of feed and grass that can be produced. It's not enough. Another problem is that there is no competition in the feed and grass industry as there's only one company. If we had our own cooperative or associated business for producing feed, things would be better.

An example of a household whose livelihood has changed little – Case Study 3

Informant: Female in her 30s

It's more than two years since I came here. I used to have more than 400 sheep and more than 10 head of cattle. When I left the grasslands I sold some animals, and the remaining 200 sheep are with my parents-in-law and brother-in-law. They are living together with another household, from whom they also rent some grazing land, and they manage the sheep together with the household's livestock. My only son goes to primary school back in our home village.

The price of feed here is very high – more than 0.31 yuan per kg. If a milk station is built near our old village I will go back as soon as possible. Everybody here feels pretty much the same way. The only good thing about being here is that it's convenient for selling milk. Since there's no milk station near our home village, I couldn't sell milk there, so I couldn't earn any money. That's the only reason I came here. The conditions are better where we were because there is grass there. And the family is separated now, which means added expenses. If it wasn't for the drought we could have managed somehow.

Since I came here, I've bought three more cows, adding to the two the government gave us. With the addition of newly born calves I had seven cows, and now I've built up to 14 cows – seven of those are dairy cows. By next year I should have 12 dairy cows. I've been lucky, because many of the calves born here were female. However, it costs a total of 10,000 yuan to raise a calf up to be a dairy cow.

At present I'm getting milk from six of my cows. Some will stop producing milk soon, while others have just started. On average I get about 120 to 125 kg of milk per day. But while the price of grass and feed has been going up, the buying price of the milk has stayed the same, so expenses are high.

I'm not used to the dairy cows yet. The other day when I washed the udder of one of the cows, it had an allergic reaction to the detergent. I was very worried.

Considering "ecological migration" as otor

From the three case studies presented above, we can see that the lives of the herders have changed with migration. From the accounts given, the following conclusions can be drawn: The previous livelihoods of the inhabitants were on the verge of ruin, due mainly to a deteriorating ecological environment, and migration was voluntary, with the initiative and support of the government. The people sold some of the livestock they possessed, but as far as possible they retained some animals in their old grazing lands. At present, although the migrants have only a small number of dairy cows, and lack experience in raising them, they are generally positive about modern dairy farming. At the same time, they hope to return to their home villages, because unlike in the migrant village, where feed costs are high, they can graze there naturally, thereby saving on the cost of feed.

The reason the herders migrated "voluntarily" is that they consider "ecological migration" as a form of *otor*. This term refers to a traditional behavior of moving temporarily to other grasslands in order to avoid natural disasters or difficult conditions. Although *otor* is essentially a voluntary phenomenon, there is also a hidden aspect of resignation or inevitability. Thus, traditionally, the question of whether *otor* is voluntary does not arise. Under the current circumstances of ecological migration, it is held that the herders themselves brought about environmental degradation, so while the economic support of the government is apparent, the coercive nature of the policy is hidden. For this reason, this form of ecological migration appears to be voluntary.

Considering ecological migration as an opportunity

The case studies presented above represent examples of migration organized by the government, but there were also cases of "voluntary" migration. In these cases, the migrants, who had experience in non-livestock-related work in their previous areas, saw "ecological migration" as an opportunity, choosing to migrate principally for the purpose of increasing their income. After migrating, they engaged in full-time livestock raising. For these people, the sense of *otor* is relatively weak, but they still intend to return to their grasslands when they have earned enough money.

An electrical appliance repairman who took a chance – case study 4

Informants: A couple in their 40s. Before migration, the husband was managing an electrical appliance repair shop in the center of the banner. The couple has two daughters. One works as a teacher, after graduating from a teachers' college; the other manages a kiosk.

We came here because of the government initiative, but also because we wanted to. At first we wanted to move to a new migrant village built by the banner government, but after thinking it over we came here. The only good thing about the migrant village built by the banner is that the buildings are clean, but there is no market to sell milk.

In this village, we started by living in the apartment building, but later, for convenience, we moved into one of the single-story houses. When we arrived, the Agricultural Development Bank of Hohhot paid a total of 3,600 yuan to cover three years of rent for our house and livestock sheds, but after moving into the house we had to pay an additional 1,000 yuan of our own money.

The two cows we got from the government gave birth to two female calves, so now we have four cows producing milk. With the birth of more calves and by buying more animals, next year we will have seven milk cows. Unfortunately, one of our cows died recently because of our lack of experience in managing milking cows.

Some of our cows produce about 20 kg of milk a day, while others can produce more than 30 kg, and one milk cow we bought ourselves is producing up to 35 kg of milk a day. But the peak milking period is only three months long, so on average we produce a total of 55 kg of milk a day, for which we get 95 yuan. Subtracting the cost of feed and grass, we can earn about 40 yuan a day. But costs can rise when cows get sick.

It's more than two years now since we moved here. At present, we are making a decent living, and maybe we are slowly reaching a level that could be called well-off. This kind of place is suitable for hard-working people. The production contract system introduced in the 1980s is good for hard-working people, but not so good for lazy people. This logic is the same everywhere. We got rid of all the livestock we had seven or eight years ago. Soon after we arrived here we were struggling.

On arriving here, we found it difficult. We had no tools or utensils, and felt quite worried. We couldn't decide if we wanted to stay here or go somewhere else. But by winter we had furniture in the house, and we felt a little more secure. At that time we didn't know anybody here, so we couldn't borrow things from anyone. And there was no means of transport. So, although we had money, we couldn't do any shopping. We could only get around after bringing a motorbike from our old village. Now, finally, we are used to living here.

Good or bad management all depends on yourself. It's a matter of skill. With skill it's possible to produce a lot of milk. For example, when we had just arrived, the cows we got from the government got sick and nearly died, but I managed to cure them myself. Getting cows pregnant also depends on skill. Anyway, compared to our lives before, things are slightly better here. Before, we had to graze animals every day, regardless of rain or wind, but here we don't have that problem. If you have skill, and you're not lazy, you won't have a problem. The lifestyle is OK too. It's a relatively nice place and very convenient. The only thing that's not so good is that I don't like the meat, because it's from small animals.

We intend to continue living here. If you have some ability, things will be OK anywhere. Back in our village we still have our house and shed, so I could go back after I expand my stock of cows. We could make a living there just by selling calves, I think. A worldly person knows how to do things the right way. Many things were stolen from our old house, though, because nobody was looking after the place.

On the question of what caused the environment to deteriorate, at the time of the "people's commune," there were eight households in our area, raising about 400 sheep. Later, the number of people increased, as well as the number of livestock, but the area of grazing land stayed the same, so naturally the environment deteriorated. Since we came here, the grazing lands back home have improved. The grass is growing high enough now to conceal sheep.

In Xinyoujia migrant village, there are also some herders who migrated "voluntarily" from other banners. One example is a household that came here from Sunid (Sunite) Right Banner, Shilingol League. In this banner, the government is implementing a so-called "enclose and shift" strategy, under which grazing land is fenced off and herders are relocated.

Two former teachers working to pay for their children's education –
case study 5

Informants: A couple in their 40s from Sunid Right Banner, Shilingol League. The couple's two daughters are living and studying at a high school in their home village. This couple previously worked as primary school teachers in the village, but when the primary school where they worked was closed down and amalgamated with another school in the banner, they both took early retirement. Although they were each receiving 800 yuan a month as a retirement allowance, they took up dairy farming to earn money for their children's education. Although there is a milk station in the banner that buys milk from dairy farmers in the area, the cost of grass and feed was high, due to the degraded state of the environment. In addition to high operating costs, life was not very convenient. For this reason, they decided without any government support to relocate "voluntarily" to Xinyoujia migrant village. Unlike other people here, they do not qualify for "poverty assistance," so they have had to cover all their own expenses, including house and feedlot rental.

When we came here we bought four dairy cows for 40,000 yuan. Now we own five dairy cows, of which three are producing milk. Of the remaining two, one had a miscarriage, and the other became infertile. We are now getting about 45 kg of milk per day.

When we first arrived we spent a lot of money, buying all kinds of tools and equipment. Raising cows in a feedlot is expensive. The price of feed is high; for example, one kg of corn costs 3 *jiao* (0.3 yuan). On average it costs about 10 yuan per day to feed a cow.

As for future plans, now that we've left our home we intend to continue managing our business here. We have four female calves now, so there's hope for the future. When we were getting milk from four cows at the same time, we were producing 65 kg of milk per day and earning more than 100 yuan a day. So if we have eight cows producing milk, we could make as much as 8,000 yuan per month. Even if 4,000 yuan of this goes on feed and other expenses, we would be able to save 4,000 yuan and enjoy a comfortable life. Anyway, unless the operation is of a certain scale, we cannot make a living. The best way of doing things would be to increase the number of cows to seven or eight back in the grasslands and then relocate here. Then things would be easy. Raising livestock is relatively easy in the grasslands because it's possible to let the animals graze naturally.

Managing dairy cows is quite difficult, because they're different to local cows in their birth and milking periods. For milking we use a semi-mechanical, semi-manual method. Each day at milking time we lead the cows to the milking station and milk them using the equipment there. If you look at the cows around here you'll notice that their udders are in bad condition. That's because of the mechanical milking. Unlike with hand milking, the machines continue milking even when there's no milk left, which damages the udders. For this reason, when we milk we use the machine to get a certain amount of milk, and then extract the rest by hand. That's why our cows have no problems with their udders.

Originally, the ex-herder interviewed in case study 4 owned some livestock and grazing land, but several years ago he gave up on livestock raising and began working in tertiary industry. The couple went to work in a nearby town. Now, the couple are using their experience of running an electrical appliance repair shop to raise dairy cows. Thanks to this experience they produce more milk than others in the migrant village. Since their livelihood is on a relatively firm footing, they have confidence both in their economic future and the improvement of the environment in their home village. Furthermore, if they are not successful in their business, they are prepared to return to their home village at any time and start raising livestock there again. In this respect, like other households, they regard their ecological migration as a form of *otor*.

The couple from case study 5 are teachers who took early retirement. Normally they would be able to subsist on their monthly retirement allowance, but to earn money for their children's education, they voluntarily joined the queue for ecological migration. As in case study 4, their attitudes stem from non-livestock-related work experience. The couple, who both worked in education – a field far removed from livestock raising – are consciously utilizing aspects of traditional livestock raising culture. This explains why they were able to devise creative ideas, such as raising a good number of dairy cows before migrating, and "semi-mechanical, semi-manual" milking. Traditional methods and modern methods are connected to a certain degree. At the same time, these refinements are the fruit of repeated trial and error in the process of implementing ecological migration.

Apart from the case studies in dairy farming introduced here, there is also a unique example of ecological migration that I will present as case study 6. In this case, the family came from the same banner as the other migrants (Xianghuang), but unlike the others, they are raising animals of a local breed – Mongolian cows. While this family's method of raising cows is exceptional, their basic mentality is typical of migrants.

Ecological migrants raising native breed cows – case study 6

Informant: Male in his 40s

I came here from Xianghuang banner. We were the only family from my village to move here.

Unlike the other people here, I only keep Mongolian cows. When I arrived I had six cows, but now I have 13. With these cows I can subsist, and even save a little. The reason I made this choice is that the local cows are adapted to the environment here, they are strong, and they don't require so much feed. But they have their downside too. They produce less milk and it's quite laborious to look after more than 10 of them. From the point of view of milk production, of course, the imported cows are good. The imported cows are also better than the local cows in terms of reproduction rate. One hundred Holsteins can produce 100 to 108 calves per year. That's not possible with Mongolian cows – they produce only one calf every two years.

In the past few years we've been keeping only a small number of livestock. This saves us a lot of trouble. Previously, we were raising more than 300 animals. But I had to sell some to pay back a debt we had for medical expenses to treat an illness.

It's now nearly three years since I arrived here. I came here completely of my own free will, but the banner government is giving me the same support as the other migrants are getting. I am grateful for that.

As for my plans, if the price of milk stays as it is, or increases a little, I will stay here for a few years and see how things go. The size of my livestock operation now is equivalent to 200 sheep. In the future, even if I sell only two cows a year, I should be able to make a living, more or less. The best thing about my situation here is that I can earn cash. This advantage will become clearer when a drought occurs.

As for the deterioration of the grazing lands, I think there are two causes. One is the lack of rainfall; the other is that the number of livestock has increased. Both the number of livestock and the number of people have been growing. The *josalang* (summer grazing land) has gradually disappeared, and the grazing lands have had no time to recuperate.

I think this policy does contribute towards the regeneration of the land. Some of the herders move away, allowing the remaining herders to use the vacated land. Both the people who leave and those who stay benefit. After some of the herders had left, the grazing land improved to an extent.

Now, in our area, spring grazing bans are being imposed on the remaining herders. This will stop the roots of the grass from being destroyed, so I think it's quite a good measure. However, if year-round bans are applied to grazing, not only will the government not get the cooperation of the herders, it won't be good for the grazing lands either. The method of keeping livestock in a shed all year round is unnatural. When animals don't move they need a lot of feed.

The thing that's worrying me now is that the price of milk is not increasing even though the price of feed is going up. If things keep going like this, I won't be able to maintain my livelihood.

The livestock raising method described in case study 6 – keeping livestock in feedlots – appears to be the government-proposed method, but in fact it is a slightly revised version of the traditional livestock raising method. This method can be continued for now, but since the native Mongolian cows are adapted to free grassland grazing, various technical issues, related to breeding for improvement and changes in mating season, are likely to arise when they are kept enclosed. However, it will take more time for these issues to come to the surface. It seems that the reason this herder has stuck with the native breed, despite knowing their disadvantages, is to leave open the possibility of returning to his old grazing land as he monitors the situation in the migrant village.

Conclusion

In the face of increasingly serious environmental deterioration in many parts of China, relocation of herders was rapidly carried out in Inner Mongolia in response to severe droughts in the first years of the new century. The main causes of environmental degradation are considered to be growth in the population of herders, and overgrazing due to increasing numbers of livestock. To address this problem, large-scale projects are being implemented at all levels of government, including the national government's "west development project." Part of these efforts involves diverting populations of herders to other industries and imposing grazing bans in grassland areas to try and separate herders from their traditional modes of livelihood. As a result, herders have been forced either to relocate under the "ecological migration" scheme, giving up their traditional subsistence patterns, or migrate voluntarily to pursue better economic prospects.

Since the establishment of the People's Republic of China, livestock raising in Inner Mongolia has undergone several major transformations as sweeping social changes have occurred. Particularly significant changes occurred in 1958, in the "people's commune" era; in the period beginning in 1983, with the introduction of the livestock contract system; and in the period from the early 1990s, when grasslands were distributed to private owners. Now, we could say we are in a period of "ecological migration," characterized by complete and partial grazing bans. Originally, it was considered that traditional livestock grazing in Mongolia was inefficient (e.g. Wang, 2000: 25–26; Liu, 2002: 47–42; Zhang et al., 2002: 4–7). With these transformations, the element of cash has gradually emerged in traditional subsistence patterns. Before the age of "ecological migration", self-sufficiency and the use of cash coexisted in livestock raising, but now, feedlot livestock raising and dairy farming are very much cash-based operations. It is clear that traditional livestock raising is facing another dramatic transformation.

Although the migrants in the case studies presented in this paper came to dairy farming in different ways, they are all following the model of economic unification. At the same time, all of them wish to return to their traditional subsistence pattern. Although they are still somewhat at a loss in their changed situation, they appear to be preserving their traditions in new ways, by seeking to balance the forces of change and tradition. It seems that this is the way in which they can sustain their traditional culture into the future.

References

Fang Zhou, 2003, Xianlü de shengtaiyimin lu, *Caoyuan Shuiwu*, Vol. 5, pp. 4–6. {Chinese}
(Fang Zhou, 2003, The manner of ecological migration living with nature, *Tax in Grasslands*, Vol. 5, pp. 4–6.)

Hou Dongmin, 2002, Caoyuan renkou shengtai yali chixu zengchang taishi yu jiejue fangfa, *Zhongguo Renkou Kexue*, No. 4, pp. 63–69. {Chinese}
(Hou DongRen, 2002, Increasing Pressure of Population Growth upon Grassland Ecology in China: Calls for ecological migration action, *Chinese Journal of Population Science*, No. 4, pp. 63–69.)

Liu Xuemin, 2002, Xibei defang shengtaiyimin de xiaoguo yu wenti tan tao, *Zhongguo Nongcun Jingji*, Vol. 4, pp. 47–52. {Chinese}
(Liu XueMin, 2002, Research on effect of ecological migration in the northwest, *Chinese Rural Economy*, Vol. 4, 47–52.)

Wang Peisheng, 2000, Shengtaiyimin – Xiaochengzhen jianshe yu xibu fazhan. *Guotu Jingji*, Vol. 6, pp. 25–26. {Chinese}
(Wang Peisheng, 2000, Construction of small cities and development in the west, *National Territory Economy*, Vol. 6, pp. 25–26.)

Wuligeng, 2003, Shilun shengtaiyimin gongzuo zhong de minzu wenti, *Neimenggu Shehui Kexue (Hanwenban)*, Vol. 24 No. 4, pp. 12–14. {Chinese}
(Wuligeng, 2003, An essay concerning ethnic problems in practice of ecological migration, *Inner Mongolia Social Sciences (Chinese Edition)*, Vol. 24 No. 4, pp. 12–14.)

Zhang Lizhong, Pan Jianwei & Sun Guoquan, 2002, Caoyuan shengtai huanjing baohu yu mumin shengcun fangshi de zhuanbian – Sunite youqi shishi "Weifeng zhanyi" zhanlüe diaochayanjiu, *Neimenggu Nongye Daxue Xuebao (Shehui Kexue Ban)*, No. 3, pp. 4–7. {Chinese}
(Zhang Lizhong, Pan Jianwei & Sun Guoquan, 2002, Environmental protection of grassland ecology and the change of herdsman's the living way, *Journal of Inner Mongolia Agricultural University, Social Science Edition*, No. 3, pp. 4–7.)

Villagers' perception of nature in relation to "ecological migration"

A case study of "A" village, Sunan Yogor Autonomous County, Gansu Province

SHINJILT

Introduction

In this chapter, I provide an overview of several of the problems faced by the inhabitants of village "A" on a daily basis due to implementation of the "ecological migration" policy. The discussion presented here is based on surveys conducted in the Sunan Yogor Autonomous County. I then extract the views of the villagers on "ecological migration" from their accounts of these problems and highlight the local development history that has shaped their views, and discuss the logic with which the local inhabitants perceive nature. Then, by looking at how the inhabitants have attempted to preserve this logic, I explore the possibilities of coexistence between humans and nature within the context of this local logic.

Rumors in the survey site

An outline of the survey site

Village A falls under the jurisdiction of the Sidalong Forestry Center *(linchang)*, in the Qilian Mountain Nature Reserve. The Qilian Mountains Nature Reserve, formally known as the "Gansu Qilian Mountains National Nature Reserve," lies between the Silk Road and the Tibetan Plateau. It is bordered in the west by the Shiyouhe River in Yumen City and by the Yongdeng Liancheng Nature Reserve in the east. More precisely, it is located between 36°43' to 39°36' North and 97°25' to 103°46' East. The nature reserve encompasses several counties and districts, including the Tianzhu Tibetan Autonomous County, the Sunan Yogor Autonomous County, Gulang, Liangzhou, Shandan, Minle, Ganzhou, and Yongchang. The total area of the reserve is 2,653,023 ha (Li Bochun et al., 2003: 33; Gansusheng Bajie Renmin Daibiao Dahui Changwu Weiyuanhui, 1997)

The nature reserve is managed by the "Gansu Qilian Mountains National Nature Reserve Conservancy," an organization affiliated to the provincial administration's forestry department. Twenty-two administrative stations fall under the administration of the conservancy, all of which receive guidance from both the local governments in the reserve as well as from the conservancy itself (Gansusheng Bajie Renmin Daibiao Dahui Changwu Weiyuanhui, 1997; An Jinling, 2002). The purpose of the reserve is to protect forests and wildlife.

The reserve is divided into areas that have been classified in three ways: "core areas" (total 72,560 ha), "trial areas" (total 390,000 ha), and "commercial areas" (total 2.194 million ha). According to Article 12 of the "Gansu Qilian Mountains National Nature Reserve Management Ordinance," in the core area, "The ecosystem is preserved in its natural state, and there are many rare species of fauna and flora on the verge of extinction." For this reason, "All organizations and individuals are prohibited from entering these core areas," and, "The governments with jurisdiction in the applicable areas must, systematically and in stages, relocate the people living within them" (Gansusheng Bajie Renmin Daibiao Dahui Changwu Weiyuanhui, 1997; An, 2002)

One of the core areas in the reserve is the Sidalong Forestry Center located in the Kangle district of Sunan Yogor Autonomous County. The area is located between 38°14' to 38°44' North and 99°31' and 100°15' East near the lower upper reaches of the Heihe River where the climate can be classified as "highlands, cold, and semi-arid". The total area of the forestry center is 175,000 ha, of which 26.3 percent is forest. The total volume of wood in the forest is estimated to be 2,274,700 m^3. The annual average temperature is 0.7°C, with average temperatures in January and July of -12.9°C and 12.2°C, respectively. Average annual precipitation is 433.5 mm, fluctuating between 326.4 and 539.7 mm. There are more than 500 species of rare plants in this area, including Qinghai spruce *(Picea crassifolia)* and Qilian juniper *(Sabina przewalskii)*. The area is also home to 230 species of fauna and contains abundant gold and coal reserves (Zhang et al., 2002; Chang et al., 2002; Chen Gang, 2002)

Rumors of ecological migration

With its many streams that feed the Heihe River and its extensive virgin forests, village A is blessed with a rich natural environment. The majority of the village inhabitants are Yogors engaged in raising livestock. It said that between the late 1950s and early 1960s, particularly during the so-called "three-year natural disaster," large numbers of Han Chinese peasants, driven by the threat of starvation, came to the Sunan Yogor Autonomous County, which was then home to many wild animals. Han Chinese arrived in village A mostly from Minle County. When the Han first arrived in the village, the Yogor herders assisted them by giving them food and also in other ways. Later, the Han began helping the Yogors with herding and also by doing carpentry work. In 1983 there were approximately a dozen Han Chinese households in the village, some of whom returned to their home villages in the same year. In 2004, there were a total of 57 households in Village A, 46 of which were Yogor, 9 were Han Chinese, and 2 were Tibetan.

I conducted a survey in the village in February 2004 using the home of Mr. "H" as the main base from which I conducted the survey (Shinjilt, 2004). With the guidance of Mr. H and his relatives, I traveled around the village on horseback to interview 10 households. The main focus of these surveys was the modern history of the area, as recounted by the inhabitants, and the people's perceptions of nature, as revealed by legends, taboos, and other cultural phenomena. However, despite my interest in these matters, the villagers often switched the topic to "ecological migration" as their primary concern was whether or not they would be relocated. They were anxious about the possibility that they would have to move from their homes in the forests of the Qilian Mountains to a "development district" *(Kaifaqu)* on the plains where they would be forced to abandon the pastoral ways of their ancestors for that of agriculture.

A typical speculation of many was as follows: "The way things are going, in less than five years time, we will be forced to vacate this area. The government is pushing ecological protection, forest protection, and wildlife protection, so they will definitely relocate us." Some of the villagers who had lost all hope in the future vented their discontent as follows: "We herders have been living with livestock for generations. But now we are told we have to become crop farmers, traders or laborers. Telling this to people like us who've never even planted fodder is unreasonable."

Yet, there was absolutely no evidence of any documents or promulgations by local governments officials stating that the villagers were to be relocated in the near future. In this sense, the people of village A lacked an "objective basis" for warranting such concerns. However, while these might only be rumors, it is true that migration has already been initiated in other areas of the Autonomous County and that several "agricultural development districts" have already been established. Given these events happening around them, coupled with the fact that the forest area they live in is already classified as a "core area" of a nature reserve, it is likely that the speculation among the villagers was more than mere conjecture.

These anxieties of the people of in village A regarding the policy of ecological migration were not only reflected by the inhabitants' behavior who switched the topic to migration without any prompting from me during my interviews, but also by the scenery around the village. Many of the herders only adopted a lifestyle centered on fixed abodes in the 1960s. For a more sedentary lifestyle it is necessary to build a permanent dwelling, yet many of the villagers had not yet undertaken to build themselves a proper house. The majority of villagers lived in decrepit houses built 30 to 40 years ago when the villagers first began settling. Given the inaccessibility of village A, transporting building materials to the area is very expensive. Consequently, anyone intending to build a house would have to be prepared to spend a lot of money. Fearing that they may soon be forced to relocate, many villagers could not make any long-term plans and were unprepared to commit to building a new house.

Forests seen as object

The conquered forests

The policy of promoting "ecological migration" in grazing areas, which is causing so much anxiety amongst herders, is based on the premise that relocating herders to other areas can alleviate the pressure on forests caused by overgrazing. According to the people of village A, however, the degradation of the forest is not their fault, but rather the result of the large-scale, organized deforestation of the area since the 1950s.

In fact, although the Sidalong Forestry Center is now within the limits of the reserve, the area has served as a timber logging site longer history as. The forestry center was established in 1956 (Chen Gang, 2002), and although logging began immediately, it was only in the 1970s when a dedicated road was opened between the forestry center and Zhangye, that the harvesting of timber really accelerated. Many of the older villagers recalled that each day from 1971 to 1978 more than 100 fully loaded trucks carried timber out of the area. One old man, Lobuzangzaoba, described the deforestation in the 1950s as follows.

> [...] In 1958, on orders from above, I climbed up the mountain with other Yogors living in the mountains and people from farming villages, and worked to cut down virgin forest. We carried the cut logs up to the snow-covered mountain using the yaks of the Yogors and burned them. We were told by the authorities that the purpose was to melt snow and let it flow into the river, to help irrigate farmland and to help in building dams [...] (Tiemuer, 2004).

Given that this corresponds to the time of " The Great Leap Forward " these accounts by the villagers regarding deforestation can be considered quite reliable. Apart from the memories of the villagers, evidence of the deforestation in Village A can also be found in academic papers. Two examples are papers about avalanches and insects.

In August 1972, an avalanche of earth and rocks of up to eight meters in diameter occurred in Tianlaochi (*Shike noor* in Yogor) Gorge, near the administrative office of the Sidalong Forestry Center. Hundreds of tons of giant rocks and trees in excess of 10 meters long were transported by the avalanche, filling up the entire gorge over an area of 13.3 km^2 (Wang Jingrong, 1983). While the cause of the avalanche cannot be directly attributed to deforestation, the following description of the damage caused by the disaster suggests that large-scale deforestation had already taken place.

> With this avalanche, estimated to have moved one million cubic meters of rocks and earth, 300 tons of coal flowed down the gorge, three dormitory buildings for forestry workers at the mouth of the gorge were destroyed, four buildings of a timber processing plant were swallowed, 14 buildings of a public health facility were buried (of which 11 were destroyed), and 3,500 m^3 of raw timber that had been placed on the river bank were carried away [...] (Wang Jingrong, 1983).

Another research paper mentions a method for preventing the multiplication of bark beetles in a large quantity of timber that had been cut down but not yet transported:

> Sidalong Forest Center is at present the largest area of virgin Qinghai spruce forest in the Qilian Mountains. Logging (management selection cutting) was conducted in this area between 1970 and 1974. However, due to the difficulty of transport and poor management of the logging site, part of the felled timber has been left on the ground for many years. As a result, there is a severe infestation of clothes moth that is placing this watershed conservation forest in a critical condition (Fu Huien et al., 1984).

Furthermore, in late 2001 a report on the Qilian Mountains Nature Reserve appeared in the "Gansu Daily" newspaper (Hou Yu et al., 2001). In this article, Liu Xiande, a former director of the "Zhangye Qilian Mountains water conservation forest research institute" discusses the causal connection between a drought occurring in Zhangye, the largest city in the region, and the deforestation in the region around the Sidalong Forestry Center. "Apart from this issue of global warming, our greedy exploitation of the Qilian Mountains has resulted in extensive destruction of our forest resources and we are thus now facing the inevitable revenge of Mother Nature." The article continues: "In the 1960s and 1970s, in line with the slogans, "more agriculture *(nongye shangshan)*" and "less forestry *(linye xiashan)*," a lot of pines, oaks and other trees were cut down." As a result, between the early 1950s and 1980 a total of 216,900 hectares of watershed forest was lost." Of particular interest is the article's description of the state, at that time, of the Sidalong Forestry Center, which was to become a core component of the Qilian Mountains Nature Reserve area.

> In the 1960s and 1970s, one after another, forestry companies set up bases in Sidalong and vigorously pursued timber processing. The buzzing of machine saws was uninter-

rupted, and the lines of tractors endless. The best hospital in the area was relocated to Sidalong. Now, after many years, the forest has shrunk, there is little timber left, the companies have closed down, the hospital has withdrawn, and the people have gone away (Hou Yu et al., 2001).

The birth and the ups and downs of forestry in Gansu Province, within which the Sidalong Forestry Center in the Qilian Mountains is located, is closely interconnected with the socialist revolutions in China and the country's national development. As Wang Shunyan of the Gansu Forestry Department says, "The forestry industry in Gansu Province was launched during 'The Great Leap Forward' in the 1950s" (Wang Shunyan, 1994: 54). It was reported that the Gansu forestry industry was previously a massive supplier of the country's timber, which was used "in order to establish socialism in fields such as industry, agriculture, national defense, science and technology, culture and education." For the 50 years up to 1999, for example, Gansu Province was the cumulative source of 26 million m^3 of timber" (Benkan bianjibu, 1999). From the time the Sidalong Forestry Center was established until the 1990s, it provided "85,000 m^3 of timber and 100 million yuan of tax revenue to the country" (Chen Gang, 2002).

As a result of the massive deforestation, "between the early 1950s and 1988, the area of forest in the Qilian Mountains fell from 5.19 million *mu* to 2.128 million *mu*, a decrease of approximately 60 percent. The proportion of forest cover in the entire area also decreased from 22.4 percent in the early 1950s to 14.4 percent" (Li Tie, 1998). In this way, as development and construction progressed, the forest was conquered and destroyed.

The protected forest

As the efficiency of logging increased, as a result of improved technology, the scale of logging in Gansu Province expanded to the extent that it was perceived as a problem. Thus, since the 1980s the scale of logging has slowly fallen. From the early 1980s, various policies concerning the protection of watershed conservation forests of the Qilian Mountains began to be implemented in Gansu Province. In 1987 the Qilian Mountains were designated a nature reserve at the provincial level and in 1998 the mountains became a national nature reserve (Li Xiaolin et al., 1998).

As a result, the 22 national forestry centers in the Qilian Mountains, including Sidalong, were given the status of "reserve administrative stations."

In 1998, following a decision by the national government to stop harvesting natural forests at the upper reaches of the Changjiang and Huanghe rivers, on October 1, 1998, Gansu Province initiated a project to protect its natural forests (Guo Lihua et al., 1999). Thus, Sidalong Forestry Center, which had become a "reserve administrative station," sealed off the mountains and began cultivating saplings, in order to contribute to a national government-proposed "natural forest protection project." In recent years, the station has been providing over 1 million saplings per year for use outside the center (Chen Gang, 2002).

The change of status from forestry center to administrative station symbolizes a shift from the era of "conquering the forest" to one of "protecting the forest". The people most directly affected by this major transition are those living within the forests. This is because, in this era of forest protection, they are positioned within the logical framework of ecological migration in such a way that they have to leave the forest in order to protect it.

In the case of the Sidalong Forestry Center, located in the Qilian Mountains at the upper reaches of the Heihe River, the logic of "ecological migration" was derived as presented below.

(1) For the past several decades, the ecosystem of the Qilian Mountains has deteriorated, causing glaciers to recede, the snow line to rise, and reducing the effectiveness of watershed forests. Thus, the volume of water flowing from the Qilian Mountains has fallen.

(2) This lead to a reduced flow of water in the main course of the Heihe River, resulting in lower groundwater levels at the middle reaches of the river and the disappearance of rivers, lakes, and marshes at the lower reaches, culminating finally in the phenomena of desertification and yellow sand.

(3) There are various causes of this problem, but the most important one is human activities in the watershed forests of the Qilian Mountains.

(4) Of these activities, the biggest problem is livestock grazing in the core areas of the natural forests, so the key to resolving these to resolve the conflict between forests and grazing *(linmu maodun)*.

(5) The most effective method of resolving this conflict between forests and grazing is "mountain seal-off for forest regeneration *(fengshan yulin)*".

(6) In order to apply "mountain seal-off for forest regeneration" "grazing discontinuation for forest restoration *(tuimu huanlin)*" and "ecological migration" must be implemented.

(7) It could be said that it was under the following three conditions that the time appeared suitable for implementing "ecological migration." (a) The national government issued "Official Notification (2001) No. 10," legislating the implementation of ecological migration of inhabitants of ecologically important districts; (b) As stated in article 12 of the "Gansu Qilian Mountains National Nature Reserve Management Ordinance," Gansu Province decided to urgently move the inhabitants within the core areas of the Qilian Mountains Nature Reserve outside the reserve, in order to completely close off the core areas; (c) In Zhangye city, the conditions to put "ecological migration" into effect are already in place. In the 1990s, 10 agricultural development districts were constructed, including Luotuocheng in Gaotai County, Xusanwan in the Sunan Yogor Autonomous County, and Shigangdun in Zhangye City. These facilities are now ready to accept ecological migrations (Chen Daqing et al., 2003; Wang Duoyao, 2001: 53–54; Chen Jinyuan, 2000: 236; Zhao et al., 2004: 207–208).

Forests seen as taboo

Villagers and nature

The logic of promoting "ecological migration" focuses on how to save the forest that has been degraded in the past, without paying any real attention to why and by whom the forest was degraded. As described in the previous section, the villager who claimed that outsiders destroyed the forests was not wrong. In an angry tone, one villager told me why the forest was destroyed, and described the impact of this destruction on the entire natural environment and on the lives of the villagers, as follows:

> If the Forestry Center and the road to the center had not existed, the forests would have remained as they had always been and we would have lived a rich life. But now that the forest has been reduced, landslides occur and precipitation has dropped; because there is less rainfall, the grass has withered. Now, we are even required to pay

> for "dry trees" (dead trees). So, we said to the people at the forestry center that if this was the case, why doesn't the center, which cut down huge numbers of "wet trees" (living trees), take responsibility. They said that they were just following the national government's instructions at that time.

Recalling the state of the forest before the 1950s when large-scale deforestation began, some of the older villagers explained the cause and effect relationship between forest destruction and the reduction in wild animals.

> At that time there were many wild animals in these mountains. Many carnivores, including bears and wolves, appeared frequently, which meant that locals strictly avoided going out alone after dark. And, because the trees were so thick in the forest, it happened that people would get lost and disappear. Now, the mountain is bald, so nobody gets lost anymore, but there are no more deer and wild yaks around.

In the logic of "ecological migration," herders are positioned in opposition to nature. Yet the people of Village A, who are considered a menace to the natural environment, insist that the destruction of the forests cannot be blamed on herders. The villagers explain how, on the basis of their customs, they could not possibly be the source of the forest degradation.

> For us, people live only for 100 years more or less, but trees can live for hundreds of years. We can't interfere with anything that lives for such a long time.

For Yogors, trees are not only an object of veneration, but they are also recognized as indicators of the future of all Yogors. For example, in Village A the following legend is told. In *Tsagan dabagan* (meaning "white peak") west of the Sidalong River, there was once a very tall *Toson hargai* (red pine) tree. Also, on a mountain known as Huamugou oboo, east of the river, there was once a *Shike noor* (Big Lake). Although *Tsagan dabagan* and *Shike noor* were several kilometers apart. Despite this fact, it is said that an inverted reflection of the tree appeared in the lake. This legend tells us just how huge the tree must have been. I was told that for as long as this miracle-like scene could be seen, the Yogors lived a very happy life. Many of the older people in the village described the natural environment many years ago, mentioning that birds such as the *galuu* would come to the lake to lay their eggs. It is believed that in those days the Yogors were very rich. To illustrate, it said that the Yogor girls at that time would make balls out of butter, now a precious commodity, just to enjoy throwing them at each other, in the manner of a snowball fight. That's how rich they were!

Villagers' perception of nature in relation to "ecological migration"

Photo 10-1. The site where the household of Mr. "H" draws their water

What we can see at the front side of the photo is a spring owned by Mr. H household of *Tsagan dabagan* settlement in "A" village. On the right side is the forest. It is said that *Toson hargai*, the tree of the Yogor legend, grew in this forest. The *Shike noor* near the upper region of the gorge that can be seen in the distance. The various taboos held by the people of Village "A" regarding nature are not limited to trees. The villagers afford equal consideration to water, especially springs. For example, it is forbidden to draw water directly from the source of a spring. On this day too, Mr. H (out of frame left) drew water from a point more than 20 meters from the source of the spring, after breaking the ice on the surface.

However, one day – nobody knows exactly when – an "evil person" cut down the *Toson hargai* tree. As a result, water began leaking out of *Shike noor* and the lake dried up. At around this time a Kangbu lama (a high-ranking priest) prophesied that the disintegration of the lake spelled the downfall of the Yogor people. After this, as predicted, the Yogors split up and scattered to different areas where they became poor and lost their status in the world. The villagers made several attempts to stop the leaking of water, mobilized by their leaders and elders, who reasoned that because it was the disintegration of *Shike noor* that led to the weakening of the Yogors restoring the lake might improve their situation, even if only a little. But each attempt ended in failure.

Shike noor is known in Chinese, as *Tianlaochi*. It is not clear if there is any connection between the legend regarding the lake, and the rock and earth avalanche of 1972 and deforestation mentioned earlier. However, for the people of Village A, *Shike noor* is very important – both as a legend and also in reality.

As alluded to earlier, in addition to *Shike noor*, trees are extremely important to the villagers are. The Yogors of Village A, so blessed with forests, classify all trees, regardless of species or name, into two kinds – *noitan modon* and *huurai modon*. The former can be translated directly as "wet trees," or more freely, as "living trees." In contrast, the latter corresponds to "dry trees," that is, "dead trees." The Yogors consider that a life force or spirit inhabits the "wet trees," and believe that these trees should never be cut down. For generations, the Yogors were told by their parents that cutting living trees was an offense as serious as killing a person. Thus, only "dead trees" could be used for building houses and for firewood. I also heard that in the past, there was a prescribed procedure for cutting down any relatively large tree – a Buddhist chant was recited before a village elder symbolically performed the first swing of the axe. There were also prescribed punishments for people who violated these taboos. In this sense, the Yogors anthropomorphized almost everything belonging to "nature" (outside world), as represented by trees, and negotiated with "nature" as an extension of human relationships.

Conversely, in the logic of national policies regarding forests, both in the era of "exploitation," and also in the era of "protection," the relationship between "humans" and "nature" can be presented as antithetical. Conversely, as can be interpreted from the descriptions above, in the logic of the villagers the relationship between "humans" and "nature" is not an antithetical one.

Responding to the rumors

In his essay, "My vanishing Qilian Mountains," the Yogor historian and writer Tiemuer wrote the excerpt below. Despite the fact that the causes of the impoverishment of water resources in the Qilian Mountains are "in fact, cultivation of grasslands, irrigated agriculture, deforestation, and mining developments," it has become commonplace in various official notices and reports to attribute the problem solely to "overgrazing by herders," and to place the burden of responsibility for the problem on "the herders, goats, and yaks in the mountains" (Tiemuer, 2004).

Undoubtedly, for the forest people of Village A, the current era of "protection," in which an attempt is made to care for forests, is preferable to the era of "exploitation," which signified deforestation. On the other hand, as the wave of environmental protection has been washing toward them since the 1990s, the local inhabitants have started to feel a sense of crisis. They have become anxious, wondering if they will be driven out of their homeland. If the migrants say, even with valid reasons, that they do not wish to migrate, the great logic justifying ecological migration, along with the various systems and policies supported that have put in place can no longer be ignored. Under these circumstances, the people of Village A are groping for a way of remaining where they are. Thinking long and hard about the future inevitability of "ecological migration," the villagers formulated the following proposal.

> Since the forest rangers at the forestry center are management-level employees of the national government, they won't have any problem getting work in other areas. But the only option for local people like us is to remain here. To protect the environment we will try to reduce our livestock herds by as much as possible. For example, someone with 100 sheep can reduce his flock to 50. Then these 50 sheep could be left to elderly people, who would find it difficult to suddenly abandon livestock grazing. The younger generations can engaged in forest protection-related work, and pour all their energy into environmental conservation. We understand this area much better than the forest rangers at the forestry center, so we are better suited to this kind of work.

For 46 years since the foundation of the PRC, there were no forest fires in the Qilian Mountains Nature Reserve. For this achievement, the Sunan Yogor Autonomous County, in which the Sidalong Forestry Center (a core area of the reserve) is located, was awarded the honor of being designated a "no forest fires county" by the state council of the PRC (Dai, 1997). Some people claim that this achievement was largely due to the efforts of herders (Li Xiaolin et al., 1998), as the herders did not merely cooperate passively with officials on the prevention of forest fires, but rather, were actively involved in the prevention of fires. Since the villagers have lived here for so long, they are skilled at riding horseback on steep mountain trails, and they know where to find trees that are dry and therefore easy to burn. And once a fire has broken out, they can easily determine the fastest route through the mountains. Actually, when the forest rangers patrol the forests for fire prevention it is impossible for them to cover the whole area, so they call on the herders in the forest and inquire about the situation. In a sense, the forest rangers cannot fulfill all their duties, such as fire prevention, without relying on the knowledge of the herders.

The forest rangers fully acknowledge that the herders are indispensable. Yet, they are not open to the idea of the herders becoming rangers. They explain that the recruitment of forest rangers is done through a unified examination, and that because the villagers are classified as "farmers and herders" in the family register system *(hukou)*, they are ineligible for the exam. At present, only management-level government employees or company employees can apply for the exam, so there is no way that the local herders can become rangers. In other words, unless you are from the city and have been educated in Chinese, you cannot become a forest ranger. Thus, as far as the system is concerned, the experience and knowledge of the Yogors in forest protection is not recognized.

Conclusion

This paper dealt with "rumors" of "ecological migration" that were spreading among the people of Village A in February 2004. But after a report in the "Gansu daily" on June 15, 2004, "ecological migration" in Sunan Yogor Autonomous County was no longer a matter of rumor. This published article stated that in the Qilian Mountains National Nature Reserve the Sunan Yogor Autonomous County "began an ecological migration project in March 2004, with the support of the national and provincial governments. Over three to five years, farmers and herders living in the core areas of the reserve will be relocated to plains and river districts with convenient transport and relatively good environmental conditions. This will reduce the pressure put on the forest by the people living there, and allow the forest to recuperate." The following plan was outlined in the report. "More than 4,000 inhabitants need to be relocated in the ecological migration project. The county's party committee and government will construct facilities to resettle the inhabitants in townships including Minghai, Qiantan, and Baiyin, and systematically implement relocation." In order to ensure a stable implementation of the ecological migration project, "the county government is dispatching officials to call on herders to persuade them..." (Yin, 2004). We cannot immediately state definitively that "ecological migration" has actually begun in Village A, but it is becoming increasingly likely.

Large-scale tree clearing was conducted in the Qilian area since the 1950s and the forest was "plundered" to such an extent that the ecosystem was destroyed. Since 1990, the need to protect forests and other natural resources has been recognized. This deforestation was undertaken based on the logic that sees nature as something to be "conquered" by humans – a logic in which cutting down trees and protecting them are opposing concepts. The idea that humans must protect nature is based on the concept that originally, humans and nature are distinct entities that (in most cases) are in conflict with each other. In this sense, the behaviors of "conquering" and "protecting" are based on the same principle and it is only the way become manifested that differs according to the times. In contrast, the local inhabitants consider cutting trees as offensive as killing people, as revealed by their various legends and taboos, and they possess a culture that sees humans and nature as essentially one and the same. They do not regard nature as an "enemy," nor as an "object" to be protected. A feature of these people's "perception of nature," if this expression can be applied to them, is that they do not *objectify* nature. And the result of this attitude is an ability to control human desire towards nature.

It has thus become markedly apparent that the logic of the policy and the logic of these people do not coincide. Ironically, after having lived in harmony with nature, they are about to be expelled from that nature. I harbor a certain amount of doubt as to whether relocating these people will really lead to the "protection of nature." As for the effectiveness of ecological migration, there is a need for scientific investigation, but it should not be forgotten that the accumulated wisdom of the people who have coexisted with nature in this area for many generations has served to protect the forests. Whether ecosystem conservation projects will be sustainable depends very much on whether the preservation of different cultures can really be guaranteed. In different living environments, there must always be more appropriate ways for people to live in those environments. One manifestation of these ways is the numerous traditional cultures of the various ethnic groups and their "perceptions of nature." In the power relationships of the real world, "master-servant" dichotomies invariably arise, between the central and the local levels. However, the relationship between people's different perceptions of nature, and the different belief systems that support these perceptions, must in principle be equal. If we can keep this firmly in mind and permit the diversity of perceptions and logic to coexist, we will always have a wide array of possibilities in front of us.

References

An Jinling, 2002, Qilian shanqu tianranlin baohu duice jianyi, *Gansu Linye*, Di liu qi, pp. 16–17. {Chinese}
(An Jinling, 2002, Suggestions of countermeasures for natural forest conservation of Qilian Mountains, *Forestry of Gansu*, No. 6, pp. 16–17.)
Benkan bianjibu, 1999, Lishi de fazhan Lishi de chengjiu –Gansu linye 50 nian huimou, *Gansu Linye*, Di si qi, pp. 10–11. {Chinese}
(The Editorial Office of "Forestry of Gansu", 1999, Historic development and historic achievement – Memoirs of forestry of gansu over the 50 years, *Forestry of Gansu*, No. 4, pp. 10–11.)
Chang Xuexiang, Zhao Aifen, Wang Jinye, Chang Zongqiang, Jin Bowen, 2002, Qilianshan linqu daqi jiangshui tezheng yu jiangshui de zailiu zuoyong, *Gaoyuan Xiangjuan*, Di er shi yi juan, Di san qi, pp. 274–280. {Chinese}
(Chang Xuexiang, Zhao Aifen, Wang Jinye, Chang Zongqiang, Jin Bowen, 2002, Precipitation characteristic and interception of forest in Qilian Mountain, *Plateau Meteorology*, Vol. 21, No. 3, pp. 274–280.)
Chen Daqing & Guan Xiaoying, 2003, Zhangye-shi Qilian-shan linqu shishi shengtai yimin de biyaoxing he kexingxing, *Gansu Linye*, Di si qi, pp. 7–8. {Chinese}
(Chen Daqing & Guan Xiaoying, 2003, Necessity and feasibility of the implementation of ecological migration in the forest in Qilian Mountain in Zhangye, *Forestry of Gansu*, No. 4, pp. 7–8.)
Chen Gang, 2002, Qilian shenchu hulü mang – Qianjinzhong de Zhangye diqu Sidalong linchang, *Gansu Linye*, No. 1, p. 22. {Chinese}
(Chen Gang, 2002, Forest Protection in the inland of the Qilian Mountains – The development of the Sidalong Forest in Zhangyi, *Forestry of Gansu*, No. 1, p. 22.)
Chen Jinyuan, 2000, Qilian-shan shuiyuan hanyanglin de baohu yu kechixu fazhan, *Gansu Huanjing Yanjiu yu Jiance*, Di shisan juan, Di si qi, pp. 235–238. {Chinese}
(Chen Jinyuan, 2000, Protection and sustainable development of the Qilian Mountain forest cultivating water resources, *Environmental Study and Monitoring*, vol. 13, No. 4, pp. 235–238.)
Dai Xingzhong, 1997, Qilian ziran baohuqu 46 nian wuhuozai, *Xiaofang Yuekan*, Di yi qi, p. 12. {Chinese}

(Dai Xingzhong, 1997, No forest fires over the 46 years in the natural reserve of Qilian Mountains, *Xiaofang Yuekan*, No. 1 p. 12.)

Fu Huien, Wu Biao, Ning Xuecheng, 1984, Xiaoduchong waijisu yinyou shiyan chubao, *Kunchong Yanjiu*, Di wu qi, pp. 213–215. {Chinese}
(Fu Huien, Wu Biao, Ning Xuecheng, 1984, Preliminary report on the experiments in the attraction of pheromone against bark beetle, *Research of Entomology*, No. 5 pp. 213–215.)

Gansusheng Bajie Renmin Daibiao Dahui Changwu Weiyuanhui, 1997, *Gansu Qilianshan guojiaji ziranbaohuqu guanli tiaoli.* {Chinese}
(The Standing committee of the Eighth Gansu Provincial People's Congress. 1997, *Regulations for the administration of Gansu Qilian Mountains national nature reserve conservancy.*)

Guo Lihua, Wang Shunyan, 1999, Shenshan chuang daye Linhai pu xinqu, *Gansu Linye*, Di si qi, p. 15. {Chinese}
(Guo Lihua, Wang Shunyan, 1999, Do a pioneering work in mountains, have an innovation in forest, *Forestry of Gansu*, No. 4, p. 15.)

Hou Yu, Yin Shangqing, Li Manfu, 2001, Qilian tanlü "Heihe liuyu guojiaji shengtai gongneng baohuqu" Zhuizong (Shangpian), *Gansu Ribao*, 12yue 21hao, Di yi mian. {Chinese}
(Hou Yu, Yin Shangqing, Li Manfu, 2001, A follow-up survey on 'environmental capability national preserve district in Heihe River Basin' in Deep Green Qilian Mountains, the first half, *Gansu Daily*, December 21st, the 1st page.)

Li Baichun, Bai Zhiqiang, Zhang Jianqi, Sun Xiaoxia, 2003, Qilianshan ziran baohuqu yu shequ jingji fazhan duice tantao, *Gansu Linye Keji*, Di er shi ba juan, Di san qi, pp. 33–35. {Chinese}
(Li Baichun, Bai Zhiqiang, Zhang Jianqi, Sun Xiaoxia, 2003, Discussion on the economic development of Qilianshan natural reserve and communities, *Journal of Gansu Forestry Science And Technology*, Vol. 28, No. 3, pp. 33–35.)

Li Tie, 1998, Moshi Qilianshan – Gansusheng Qilianshan shuiyuan hanyanglin baohu zongshu, *Gansu Linye*, Di yi qi, pp. 7–9. {Chinese}
(Li Tie, 1998, May not Qilian Mountains be lost – general theory of the protection of water resourses forest of Qilian Mountains in Gansu, *Forestry of Gansu*, No. 1, pp. 7–9.)

Li Xiaolin, Xiao Yuchun, 1998, Zai Hexi zaolang yu Qilian zhi jian, *Minzu Tuanjie*, Di shier qi, pp. 39–42. {Chinese}
(Li Xiaolin, Xiao Yuchun, 1998, Between the Hexi Corridor and Qilian, *Ethnic unity*, No. 12, pp. 39–42.)

Shinjilt, 2004, Kokuga jyoryuiki no hito to shizen – Seikai-sho Chiren-ken Kanshuku-sho Shukunan-ken deno chosa hokoku, Sogo Chikyu Kankyogaku Kenkyujyo Oashisu purojekuto Kenkyukai, *Oashisu Chiiki Kenkyu Kaiho*, Vol. 4, No. 1 pp. 111–130. {Japanese}
(Shinjilt, 2004, Relationship between human and nature in the upper reaches of the Heihe River: an ethnography on Qilian and Sunan counties in Qinghai and Gansu., Research Institute for Humanity and Nature / The Oasis Project, *Project Report on an Oasis-region*, Vol. 4, No. 1, pp. 111–130.)

Tiemuer, 2004, Shi wo Qilianshan, *Yanan Wenxue*, Di wu qi, pp. 160–165. {Chinese}
(Tiemuer, 2004, My vanishing Qilian Mountains, *Yanan Wenxue*, No. 5, pp. 160–165.)

Wang Duoyao, 2001, Fengshanyulin zengqiang Qilian-shan shuiyuan hanyanglin xiaoneng, *Gansu Linye Keji*, Di er liu juan, Di si qi, pp. 52–54. {Chinese}
(Wang Duoyao, 2001, Enhancing water conservation by closing hillsides to facilitate afforestation, *Journal of Gansu Forestry Science and Technology*, Vol. 26, No. 4, pp. 52–54.)

Wang Jingrong, 1983, Fasheng zai Qilian-shan beilu de yici nishiliu, *Bingchuan Dongtu*, Di wu juan, Di er qi. {Chinese}
(Wang Jingrong, 1983, A debris flow in north side of Qilian Mountain, *Journal of Glaciology and Geocryology*, No. 5, Vol. 2.)

Wang Shunyan, 1994, Guanyu Gansu linye qiye guanlizhong de jige wenti, *Zhongguo Linye Qiye*, Di yi qi, pp. 53–54. {Chinese}
(Wang Shunyan, 1994, Several problems in the management of Gansu forestry enterprise, *Chinese Forestry Enterprise*, No. 1, pp. 53–54.)

Yin Shangqing, 2004, Huanjie linqu renju yali zengqiang shuiyuan hanyang gongneng – Sunan shengtai yimin zaofu nongmumin, *Gansu Ribao*, 6yue 15hao. {Chinese}
(Yin Shangqing, 2004, Ecological migration brings great benefit to the herdsmen in Sunan county – alleviating habitat stress in forests and enhancing water conservation, *Gansu Daily*, June 15th.)

Zhang Hu, Zhang Hongbin, 2002, Qilian-shan Sidalong linqu jixue bianhua qushi yu qiwen he jiangshui, *Gansu Linye Keji*, Vol. 27, No. 2, pp. 1–4. {Chinese}
(Zhang Hu, Zhang Hongbin, 2002, Changes of snow cover and its relations to temperature and precipitation in Sidalong Forests of Qilian Mountains, *Journal of Gansu Forestry Science and Technology*, Vol. 27, No. 2, p. 1–4.)

Zhao Chengzhang, Fan Shengyue, Yin Cuiqin, 2004, Qilianshanqu tianran caodi tuihua yuanyin fenxi yu kechixu liyong duice, *Zhongguo Shamo*, Er yi si juan, Di er qi, pp. 207–210. {Chinese}
(Zhao Chengzhang, Fan Shengyue, Yin Cuiqin, 2004, Analysis on natural grassland degradation reason and its sustainable utilization countermeasures in Qilian Mountains, *Journal of Desert Research*, Vol. 24, No. 2, pp. 207–210.)

Differences in perception among the parties concerned with the "ecological migration policy": from case studies in "B" Township, Qifeng District, Sunan Yogor Autonomous County, Gansu Province

11

Tomoko Nakamura

Introduction

The "ecological migration policy" as discussed in this paper, refers to the phenomenon of forced migration that is currently being implemented in China, nominally for the purpose of environmental protection[1]. The ecological migration policy was initially put into effect in the Inner Mongolia Autonomous Region in July 2001 (People's Daily Online, Japanese version, July 29, 2001), after which the policy was applied throughout China. As is typical of the policy, in the Sunan Yogor Autonomous County – the focus of this paper – ecological migration policy involved transferring people between various places, for example, from D to E Township, or B to C Township (see Fig. 11-1). As academic researchers have commenced studying the policy, the progress of the policy's implementation and field surveys have started to be reported at academic conferences and the like (Mailisha, 2004; Borjigin Sergelen, 2003). Also, in April 2005, in Chicago, the annual meeting of the AAS [Association for Asian Studies] was on the theme, "Ecological Migration: Environment, Ethnicity, and Human Rights in Inner Mongolia". Such reports reveal that there is considerable debate in the places where the policy is being implemented regarding the causes for environmental degradation, and that there are various problems associated

---

1 In this paper, the term "ecological migration" is used in an emic sense as a concept in the Chinese language that is applied to phenomena currently occurring in China. When adding etic explanations that can be understood cross-culturally, as in my examination towards the end of the paper, "ecological migration" can be viewed as a form of "migration caused by a nation-wide forestation policy," and the policy that produces such migrants can be viewed as an "ecological migration policy".

with the policy, such as the declining standards of living of the migrants. Many of the social issues associated with ecological migration are discussed in the numerous reports presented in this book. The questions of how these issues have arisen, and why they have become problems need to be considered. I believe that the key to answering these questions is to pay attention to the people affected at different levels of the policy implementation process, and to resolve these issues, one by one, at their various levels.

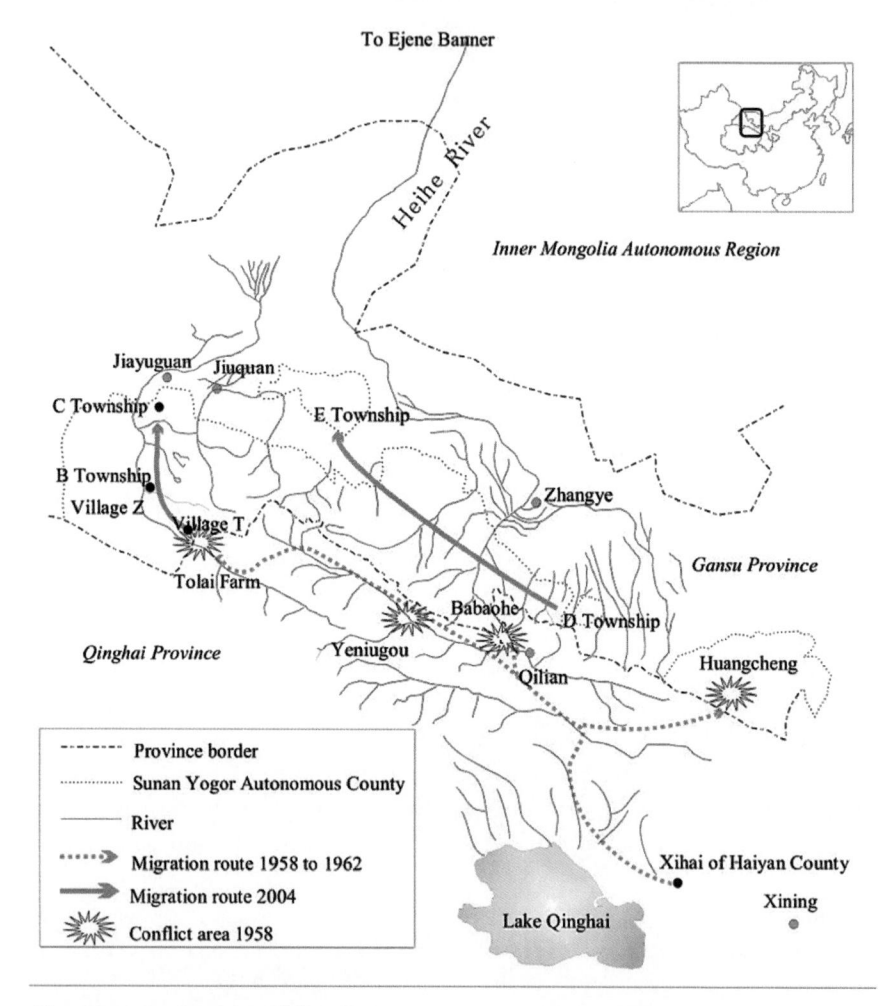

Fig. 11-1. Outline map of the survey area
Note: Created by Nakamura Tomoko

Many reports relating to these issues were presented at an international symposium entitled "Environmental Immigration – Practice and Experience" held in Beijing in July 2004. Many of the case studies presented at this symposium presupposed a scheme in which the government and local people were set against each other in regard to the policy, as "subject" and "object" in an antithetical relationship. However, it is not necessarily true that this presupposition carries through from the government to all the persons directly affected by the national policy. In order to attain a more realistic understanding of the phenomenon of ecological migration, it is necessary to abandon this presupposition, and to examine the processes of policy formulation, communication and implementation. It is also necessary to discover the reality of what happens during each of these stages, to analyze the "viewpoints" of the people concerned at each stage, and to clarify the mechanisms of the social phenomena associated with ecological migration. Here, "viewpoint" refers to the aspects of the policy about which the people affected are most concerned. This report focuses on the differences in these "viewpoints" at the various levels of the ecological migration process, and aims to clarify the real effect of national policy on local people, and the mechanisms behind these effects.

Background on the implementation of ecological migration policy

To start with, in order to provide some background on the implementation of ecological migration, I will present a brief description of the area considered in this study.

The township considered in this study belongs to the Sunan Yogor Autonomous County in Gansu Province and is located at the northern foot of the Qilian Mountains, which rise more than 4,000 m above sea level. Although most of the mountains are covered by shrubs, forests of Quinghai cedar can be found at various places below the tree line at 3,500 m. For this reason, the region plays an important role as a water recharging area for the Heihe river system (Pan and Tian, 2001).

Water utilization of the Heihe River is currently receiving considerable attention. It has been pointed out that the use of water for irrigation along the middle reaches of the river has influenced the livelihoods of herders in the lower reaches – in Ejene Banner, Alasha League and in the Inner Mongolia Autonomous Region. Since the 1970s, the rivers in Ejene Banner have been drying up in the summer, adversely affecting the lives of herders. For example, the herders have almost completely ceased raising cattle and horses (Ozaki and Nakamura, 2002). Unable to ignore this situation, on December 1, 1981, the government of the Inner Mongolia Autonomous Region raised the issue of unfair utilization of water resources with the state council. Since that time, the issue of water utilization along the Heihe River has been discussed on numerous occasions by the state council. Later, in the early 1990s, the move to conserve the water resources of the Heihe River gained momentum. With this goal in mind, the need to protect the forests in the river's catchment area was emphasized[2]. At this time, two problems were identified in the upper reaches: deforestation that occurred in the 1980s in the Qilian Mountains, and the reduction in the amount of forest as a result of overgrazing by mountain herders. The cause of the latter problem was identified as being the goats kept by mountain herders eating the shrub forests. The solution adopted for this problem was the policy of "cultivation discontinuation for forest restoration", described in detail in the next section. A concrete aspect of this policy was the initiative of ecologic migration, namely the relocation of the mountain herders. Under such circumstances, beginning in July 2004, ecological migration in "B" Township – the focus of this chapter – was undertaken.

I conducted aural surveys in this area in August 2003, and in February and May 2004 – before ecological migration had commenced. At the time of my survey in 2003, people had not yet received any notification from the government about the impending ecological migration initiative.

The inhabitants of the area are a mixture of various ethnicities, including Tibetans, Yogors, Muslim(Hui)s, Han Chinese, Mongolians and Tus, with Tibetans making up the majority. In 1999, a total of 3,281 people from 993 households lived in Qifeng District, of which 85.37 percent were Tibetans (Sunan Yuguzu Zizhixian Renminzhengfu, 1999: 105–107). According to statistics for 2000, 393 people from 124 households were living in "B" Township and the majority of these, with the exception of

---

[2] Since 2000, water supply controls have been effected along the Heihe River, as well as on the Huanghe and Tarim Rivers, under the guidance of the National State Council (Pan and Tian, ed. 2001: 117).

township officials, were engaged in livestock herding. These people made their living through "mobile livestock raising", whereby they raised groups of animals – mainly yaks and Tibetan sheep – by moving twice or three times a year to graze in mountainous areas at altitudes between 2,500 and 4,000 m. I will describe one typical example of this grazing method.

Mr. "A" (born 1955, Tibetan) works as the secretary of Village "T". He owns 100 goats and 410 Tibetan sheep – close to the average number of livestock per household in the township (547). Mr. A earns a salary of 4,900 yuan per year in his capacity of village secretary, whereas his annual earnings from selling livestock and livestock products such as hides and hair amounts to almost 70,000 yuan. This means that raising livestock is the foundation of his livelihood, with his salary providing only supplementary income. Mr. A's income may be higher than average for herders in B Township, but given that his job as secretary is not financially important, I think he can serve as a useful typical example without any problems.

The following is a generalized description of the annual cycle of "mobile livestock raising" practiced in this area, based on aural surveys of Mr. A and other leaders in B Township. Firstly, there are three kinds of campsites – for summer, autumn, and winter. On June 20 each year the herders move from their winter campsites, where they spend winter and spring, to their summer campsites located at more than 3,800 m above sea level. On August 10 each year, they move to their autumn campsites at 3,000 to 3,800 meters. On December 1 the herders move back to their winter campsites at approximately 3,000 m. The administrative center of B township is located at an altitude of about 2,900 m, and the campsites are scattered on the mountains surrounding the town.

The means of transport used by the herders are the jeeps owned by each household. All three kinds of campsites are located at the southern foot of the main range of the Qilian Mountains. Since they are situated at the southern foot and because of the extreme altitude, there are no forests, so the livestock are grazed in grasslands. Mr. A's summer and autumn campsites are adjacent to each other, so he refers to them collectively as his summer campsite, but the timing of his relocation to his summer and winter sites accords precisely to the annual cycle described earlier. Mr. A's family consists of five people. His eldest son, a schoolteacher, and his second son, a student, both live away from home. Mr. A and his wife, along with their daughter and a herder-laborer are engaged in raising livestock, moving together between the campsites. In many cases elderly people and children do not move between the campsites, spending the whole year at the winter campsite.

The grazing land in B Township is strictly divided between individual households in the area surrounding the winter campsites, but at other campsites the grazing area is roughly assigned in portions for groups of five households. According to the leaders of the township, the grazing area per household at winter campsites is approximately 4,000 to 5,000 *mu* (2.7 to 3.3 km^2). At the autumn and summer campsites, the area of the grazing land is unknown, since areas are not measured for the reason cited above. In the case of Mr. A, he has about 7,000 *mu* (4.3 km^2) at his winter campsite and 3,000 *mu* (2.0 km^2) at his summer campsite. However, his winter area includes 4,000 *mu* (2.7 km^2) of common village land that he leases. The herders of the township graze their livestock in natural grasslands, around campsites located at the highest altitudes of the Sunan Yogor Autonomous County, yet they are realizing a shortage of grazing land at their winter campsites, despite their mobile grazing method that utilizes a range of altitudes. Mr. A has 2.5 *mu* (1,600 m^2) of crop land at his winter campsite, which he uses to grow oats for livestock feed – managing harvests of about 2.5 tons. However, this is not enough to meet his minimum needs. Also, because harvests vary greatly from year to year, Mr. A normally needs to purchase several tons of feed from agricultural areas to have enough feed to get through the winter (Photo 11-1).

As already mentioned, Mr. A and other herders currently derive the bulk of their income from livestock. Here I will give a more detailed description of their income structure. According to aural surveys, Mr. A's gross income in 2003 was derived partly from the sales of 52 live goats (average price 250 yuan) and 230 Tibetan sheep (average price 200 yuan), which together amounted to earnings of 59,000 yuan. In addition, Mr. A earned 400 to 500 yuan from sheepskins, 5,700 yuan from cashmere, and 3,800 yuan from sheep's wool. From this we can ascertain that Mr. A earns the bulk of his income from sales of live animals.

Parenthetically, I noticed something strange about the composition ratio of livestock of Mr. A and other herders – the absence of yaks, which sell for about 1,000 yuan in the case of a mature animal. The yak is a typical livestock animal of the Qilian Mountain region. According to the leaders of B Township, there were 1,580 yaks in the township in 1984, before the privatization of livestock ownership. Soon after privatization they were sold off, and now only two families in the township keep yaks,

Photo 11-1. Sheds at a winter campsite

with less than 100 animals between them. The decline in the number of yaks relates to the fact that they roam over a large area and that they tend to enter the grazing areas of other people, which can create problems, particularly when they cross over into the grazing lands of Qinghai Province. In short, the yaks were sold off because of their potential to cause grazing land disputes; this aspect is discussed more fully later.

Finally, I wish to discuss the water utilization of herders in the township. Neither Mr. A nor any of the other herders irrigate their lands. This means their livestock operations are heavily dependent on natural grass resources, which in turn rely on rainwater. They also depend on rain for drinking water. At present, Mr. A uses water pumped from a 156-m-deep well located 500 m from his home at his winter campsite. The water is conveyed by pipe from the well to his home. This well was dug in 1999 in response to river water pollution concerns due to the opening of a mineral processing plant that began operation in the center of B Township in 1997. Up until that time, people were using river water directed through water channels for their drinking water. Water from these channels is now used only for irrigating crops. On the other hand, at the summer and autumn campsites drinking water is still taken directly from springs and rivers.

In this area, the ecological migration policy aims to relocate the people pursuing livelihoods as described above to semi-arid areas at an altitude of 1,800 meters (C Township, in Qifeng District, approximately 46 km south of Jiayuguan). Migration to these semi-arid areas means that the herders are forced to change to intensive livestock raising, whereby livestock are raised in feedlots and the herders need to grow feed crops. The migrant households are to be provided with a house, a livestock shed and a well, as well as 50 *mu* (3.3 ha) of cropland and 50 *mu* (3.3 ha) of grazing land. In addition, the migrant village is reportedly equipped with infrastructure that includes a school and a hospital. At the time of my survey (July 2004) the first stage of an ecological migration program to relocate 10 households (715 people) from Village "T" and Village "Z" in B Township was planned. A total of approximately 1,000 households from Qifeng District (including B Township) were to be relocated within three years. As mentioned earlier, the number of households in Qifeng District in 1999 was 993 (based on general population trends in the area, this number is likely to have increased to some extent by 2004). Thus, we can conclude that practically all the people in the district were to be relocated.

Perceptions of parties concerned and past migration policies

Now I would like to consider the viewpoints of each category of the parties directly involved in this policy – specifically, the national government, regional governments, local governments, and the inhabitants.

Firstly, what is the viewpoint at the national level, from which the policy was originally issued? As far as I have been able to determine, there are no policies that describe the ecological migration policy in detail. So, I will concentrate on the "cultivation discontinuation for forest restoration" policy, since that is the basis of the ecological migration currently being implemented in B Township, and I will attempt to clarify the underlying viewpoints behind this policy.

The first significant mention of ecological environment protection by the national government was in 1998, in the wake of the massive flooding of the Changjiang River that occurred in that year. The state council considered agricultural development in mountain areas to be one of the reasons for the devastating scale of the floods, and within the same year (1998) officially announced a national ecological environment plan. This document states that, "The ecological environment is a fundamental condition for the development of humanity", and clearly states that protection of the ecological environment will enable the sustainable development of China. It describes in detail the goals for environmental recovery to be achieved by 2050, along with the current state of particular areas recognized to be environmentally degraded, such as the Tibetan Plateau and the Huanghe River Basin, and sets environmental recovery targets for these areas for 2010. On December 21, 2000, the state council issued "guidelines for national ecological environment protection," a revision of the 1998 plan. In these guidelines, a new goal was added – environmental conservation in areas that were in relatively good environmental condition. It was stated that individual local governments and regional bodies were responsible for putting into effect concrete policies based on these guidelines. Then in March 2001, based on these guidelines, 10 areas were designated as national ecological reserves, with the national environment bureau ordering that the ecological environment in these areas be improved. In response to this, within these 10 reserves "ecological migration" was commenced on a trial basis, for the purpose of regenerating vegetation, under the slogan of "cultivation discontinuation for forest restoration." Based on the results of the trials implemented in these areas, in 2003 the "cultivation discontinuation for forest restoration ordinance" officially came into effect. As its name indicates, this ordinance aims to afforest agricultural land. The most notable points of the ordinance are as follows. The first is that the highest priority is placed on ecological protection; the second is to encourage the implementation of ecological migration as part of the forest restoration policy; the third is that local governments are charged with the responsibility of conserving the ecological environment after it has been restored. The term "ecological migration" was first used in this document.

Based on an overview of the policy flow leading to the ecological migration policy, it is apparent that at the national level "ecological environment protection" is at the forefront of the initiative. In other words, at the national level, utmost importance is put on the ecological environment.

We can infer that priority is placed on "ecological environment protection," or that it is "ecology" rather than "migration" and "relocation" that is of primary concern.

Next, we look at the local government level, which formulates and implements specific policies. In this case, we can divide local government into two levels – the province level, which formulates concrete plans, and the township level, which directly applies the policies to the inhabitants. An official document released in January 2004 by Gansu Province contains the following description of a concrete policy proposal relating to environmental protection of the upper reaches of the Heihe River; "Mountain seal-off and grazing bans shall be comprehensively implemented in natural forest areas, public forest areas, managed reforestation areas, ecologically fragile areas […] regenerated forest, as well as mountain areas sealed off for forest regeneration". That is, the viewpoint at the local government level does not seem to differ substantially from that at the national level in terms of the special importance given to forest restoration. On the other hand, the township government states that, "Being located at the southern foot of the Qilian Mountains, there were no forests originally in this area – only grasslands." [The forests are at the northern foot of the mountains.] However, the township government continues to state that, "In this sense, we cannot have forest restoration. It is a fact, however, that the environment is deteriorating, with visible evidence of ground subsidence due to the proliferation of rats, and a reduction in grass height. So relocation must be implemented." The township government goes on to say that, "Since the migration destination is close to a town [Jiayuguan] that possesses infrastructure, the standard of living will be higher, so it is best for the inhabitants to migrate." Despite the fact that the township government expresses doubts about the province's pretext of ecological protection, it is general agreement on the point of the deterioration of the natural environment; it reinterprets "watershed conservation forest recovery" as "water retention," and emphasizes "vegetation" as grassland. Furthermore, the township government adds one more incentive for migration – the inhabitants will enjoy an improved standard of living. To summarize, we can infer that in addition to "ecological environment protection," the township's viewpoint focuses on the act of "relocation" and, as a pretext for this, the resulting "economic effect."

What about the viewpoint of the local people on this kind of ecological migration?[3] They understand that the national and regional governments are claiming protection of the natural environment to justify the need to implement ecological migration. They also know by direct experience that the environment is deteriorating, as evidenced by the fact that the height of grass in the grasslands has declined from about 50 cm in the 1980s to only 10 cm, and that the density of the grass is also lower. On the other hand, they also claim that (if the purpose is to allow grass regeneration) they would be prepared to reduce their livestock holdings in order not to relocate. In addition, the people say that despite the fact that there are alternative methods for achieving environmental recovery, these alternative methods have not been considered. They provided an alternative explanation of how the grasslands became degraded in relation to their history.

According to the herders, the history of the area is a tragic one. Between June and July in 1958 large numbers of herders in Qilian County, Qinghai Province, crossed over the Heihe River into Sunan County. The origins of this relocation lay in government forcing people out of Xihai Town in Heiyan County in Quinghai Province in order to construct a nuclear energy test facility (now called "Yuanzi cheng"). These forced migrants moved north to the Heihe River, and settled near the border of Quinghai Province and Sunan Yogor Autonomous County – now the Tolai farm and Babao River area (both in Sunan County at the time) – and Huangcheng (Qinghai Province). The arrival of these migrants led to friction with the original inhabitants (Fig. 11-1). In December of the same year, a government report was issued containing proposals to try and resolve the conflict in the border region (Sunan Yuguzu Zizhixian Renminzhengfu, 1999: 451), and land was exchanged between Gansu Province and Qinghai Province. Although an attempt was made on solving the problem in 1958, migrant herders continued to arrive and friction persisted until 1963. As a result, between 1959 and 1962 almost 7,000 people on the Sunan side moved away to various other places. This event is known locally as the "great exodus."

One informant said to me that at the time of the "great exodus" the population and livestock numbers increased greatly, suggesting that one factor behind the increase in livestock numbers was this previous migra-

3   Of course, it is possible that there is a certain degree of diversity among the people categorized as "local people". I intend to examine this question in a detailed analysis in another paper.

tion policy. In fact, at the time of the "great exodus," a total of 135 migrants of 27 families arrived in Village "T" in B Township, which is now subject to the current ecological migration policy, and 144 migrants of 24 families arrived in Village "Z". In 1970, the population of Village T was 211 people in 39 households, and that of Village Z was 83 people in 36 households. Thus, we can conclude that the migration caused a large increase in population. Furthermore, in 2002, a second informant told me of a rumor that in 1983 an area west of the Maying River, east of the Shiyou River, south of Jiayuguan, and north of the headwaters of the Tolai River, was designated as the "Qilian Mountains Nature Reserve." This area overlaps with the area of the 1958 conflict. This herder related that this rumor led people to anticipate a possible second "great exodus", following on from the first one in 1958. To summarize, while the first herder saw the "great exodus" of the past as the cause of the current ecological migration, the second herder predicted the ecological migration policy – perceiving it as following on from the "great exodus" of 1958. At first glance it seems that the "great exodus" is a thing of the past. However, even now, conflict over grazing land is a serious problem in the lives of the herders. While conducting my surveys, I heard that households of herders owning large numbers of livestock from the townships close to the "capital" of the Sunan Yogor Autonomous County let their livestock graze in vacant land in B Township. If land is left vacant, herders from Qinghai Province move into the area, so to prevent this the herders and the government of B Township permit this grazing. Thus, at the level of the herders, both in the past and in the present, conflicts over grazing land are a major problem that are deeply rooted in their way of life.

To summarize, the viewpoint of local herders focuses more on "migration" and "relocation" rather than "ecological environment protection." And on considering the elements of the policy they have come to understand that the current migration is connected to migration policies of the past. Furthermore, "migration" and "relocation" is now the most essential issue in their lives, and its development will have a major bearing on their futures.

Conclusion

As I have shown, it is apparent that the parties concerned with ecological migration at the various levels of its implementation are each focusing on different aspects contained in the policy. As I have stated, the national level focuses on "ecological environment protection" that aims to achieve the regeneration of vegetation – principally forests. At the province level, the viewpoint is generally similar to that of the national level. On the other hand, instead of the national level's concern for the regeneration of forest vegetation, the township level interprets "ecological environment protection" to mean restoration of grassland vegetation. The township also sees migration as a means to achieve a higher standard of living for the inhabitants. In other words, the township government has actualized the policy emphasizing "relocation" as well as environmental protection. At the same time, it is clear that the viewpoint of local herders focuses on "migration", due to their problem with grazing land conflicts. It is reasonable to suppose that all these points of focus are interrelated with the connotation that the phrase "ecological migration policy" holds for each of the concerned parties and their values.

After systematically summing up these different viewpoints, the following can be said. In the case of B Township, while the national government views the future through its administrative system in response to natural phenomena such as floods and deforestation, herders are looking to the future through historical memory or culture, in response to social phenomena such as migration or grazing land conflict. Using the local government as intermediary, the herders and the national government are engaging in "communication based on *dis-communication*"[4]. This scheme describes precisely what is happening now in B Township in relation to ecological migration. It is possible that this structure applies to social phenomena arising from ecological migration in other areas as well.

This paper focused on the differences in the "viewpoints" at various levels relating to the ecological migration policy, and attempted to clarify the real state and structure of how the policy at the national level is

---

4 When the parties concerned realize the "gap" between their viewpoints and their perspectives, it is possible that a new dynamics will arise. I will take up analysis of this issue in a future study.

actually being implemented at the level of the people affected locally. It is important to appreciate that national policy and the sentiments of the people ultimately affected by it cannot be expressed as a simplistic formula of a direct antithetical relationship – a more complex mutual interaction is at work – due to the "gap" in viewpoints and the perspective of the parties concerned.

References

Borjigin Sergelen, 2003, Uchimongoru-jichiku no hobokukinshi to "Seitaiimin" no jittai, *Warudo Torendo*, 9 gatsugou, pp. 47–49. {Japanese}
(Borjigin Sergelen, 2003, A Prohibition against pasturage and the reality of ecological migration, *Ajiken World Trends*, September, pp. 47–49.)

Gansusheng Sunan Yuguzu zizhi xian difangzhi bianzuan weiyuanhui, 1994, *Sunan Yuguzu Zizhixian Zhi*, Gansu Minzu Chubanshe. {Chinese}
(Gansu Province Sunan Yogor Autonomous County Government, 1994, *Annals of Sunan Yogor Autonomous County*, Gansu Ethnic Publishing House.)

Haibei Zangzu Zizhizhou Diming weiyuanhui, 2001, *Haibei Zangzu Zizhizhou Dimingzhi*. {Chinese}
(Haibei Tibetan Autonomous Prefecture Place-Names Committee Office Edition, 2001, *Annals of Place-Names of Haibei Tibetan Autonomous Prefecture*.)

Jiang Xuguang bianji, 2001, *Lüse Songge – Huanghe Heihe Talimuhe Diaoshui Shilu*, Zhongguo Shuili Shuidian Chubanshe. {Chinese}
(Jiang Xuguang et al., 2001, *Paean to Green 7 The Observations on Water Control of Yellow River, Heihe River and Tarim River*, China Waterpower Press.)

Mailisa, 2004, Seibudaikaihatsu no naka no shosuminzoku seitaiimin- Shukunan Yuguzoku jichiken ni okeru chosahoukoku, Aichi Daigaku Gendai Chugokugakkai hen, "*Chugoku 21*", 18, pp. 79–86. {Japanese}
(Mailisa, 2004, Ethnic minority immigrants under the western region development: A report from the Sunan Yogor Autonomous County, Aichi University Faculty of Modern Chinese Studies, *China 21*, 18, pp. 79–86.)

Ozaki Takahiro and Nakamura Tomoko, 2002, Echina bokuchiku chosa houkoku, *Jimbungakka Ronshu*, 56, pp. 45–90. {Japanese}
(Ozaki Takahiro and Nakamura Tomoko, 2002, Investigation report on pastoralism in Ejina Banner, *Cultural Science Reports of Kagoshima University*, No. 56, pp. 45–90.)

Pan Houmin and Tianshuili bianji, 2001, *Heihe Liuyu Shuiziyuan*, Huanghe Shuili Chubanshe. {Chinese}
(Pan Houmin and Tianshuili et al., 2001, *Water Resources in the Heihe River Basin*, Yellow River Conservancy Press.)

Qifeng-Qu Gongshu, 1994, *Qifeng Zahnzu Lishi Gaikuang.* {Chinese}
   (Qifeng District Government office, 1994, *Survey of Qifeng Tibetan History.*)
Renmin Ribao, 2002. {Chinese}
   People's Daily, 2002
Sunan Yuguzu Zizhixian Renminzhengfu, 1999, *Gansusheng Sunan Yuguzu Zizhixian – Xianqing yu Kaifa.* {Chinese}
   (Sunan Yogor Autonomous County Government, 1999, *Gansu Province Sunan Yogor Autonomous County – Circumstance and Development of the County.*)
Zhonghua Renmin Gongheguo Guowuyuan, 1998, Quanguo shengtai huanjing jianshe guihua. {Chinese}
   (The State Council of the People's Republic of China, 1998, The national plan for ecological environment construction.)
Zhonghua Renmin Gongheguo Guowuyuan, 2000, Quanguo shengtai huanjing baohu gangyao. {Chinese}
   (The State Council of the People's Republic of China, 2000, The outline of the national-wide protection of ecological environment.)
Zhonghua Renmin Gongheguo Guowuyuan, 2002, Tuigenhuanlin tiaoli. {Chinese}
   (The State Council of the People's Republic of China, 2002, Regulation of returning farmland to forest.)

Conclusion

Global environmental problems and ecological migration

MASAYOSHI NAKAWO

Global environmental problems

It has been more than 30 years since the world began to take notice of global environmental problems. This began with scientists warning that the average global temperatures of the earth were rising at an abnormally fast pace, and that this was leading to accelerated melting of polar icecaps and glaciers, which would cause sea levels around the world to rise. The first identified cause of this global warming phenomenon was the now well-known "greenhouse effect" resulting from the increasing concentrations of CO_2 in the atmosphere due to the considerable global consumption of fossil fuels. Since the majority of the world's large metropolitan areas are located at less than a few meters above sea level, it would be a disaster of unprecedented proportions if sea levels were to rise. Since the earth's atmosphere is a globally shared resource that extends around the world, global warming issues and their associated environmental threats have received considerable coverage by the mass media as issues affecting the entire world. There is now considerable awareness of global warming among both policy-makers and the general public in many countries.

Triggered by these concerns over global warming, attention was subsequently shifted toward a variety of other so-called global environmental problems that were arising in different parts of the world. These issues included ocean pollution, expanding desertification, the appearance of holes in the ozone layer, and the loss of biodiversity, all of which illustrated the urgent need to tackle these problems.

The main difference between global environmental problems and ordinary pollution is that all human beings living on the earth are, concomitantly, both victims and perpetrators of the problem. In addition, various environmental problems are manifested in unexpected ways and locations that extend beyond the borders of regions and nations.

These problems are the result of considerably complicated interactions between the system of "nature" and human activities and there effects are almost impossible to predict based solely on scientific knowledge; these are the kinds of problems we are currently face. Of course, it is important to direct our efforts toward clarifying the causal relationships to try and solve these problems, but given that the problems are so difficult to foresee, I wonder if doing this alone is sufficient.

For example, if we consider the prototypical global environmental problem of global warming, nobody imagined that fossil fuel consumption would lead to such a problem. For several centuries humanity has relentlessly pursued an ever more comfortable lifestyle based on scientific and technological developments, such as the invention of the steam engine and the diesel engine. It is now understood that global warming is affecting the world's vegetation and agricultural products beyond their capacity to adapt and this has caused a variety of unforeseen problems, including the frequent occurrence of abnormal weather phenomena such as torrential rain, droughts, heat waves, and heavy snowfalls all over the world. In other words, an important attribute of global environmental problems is that human activities can lead to unintended results.

The selection of the *kanji* character by the famous Kiyomizu temple in Kyoto to symbolize the year 2004 was 災, meaning "disaster", is still fresh in my mind. The selection of this character was not surprising given that it was the year in which Japan experienced numerous natural disasters, including record-breaking typhoons, frequent flooding caused by the typhoons, and a major earthquake in the Chuetsu region of Niigata prefecture. These events were followed by the earthquake in Sumatra and the resulting massive tsunami that devastated numerous countries across the Indian Ocean. All this left the strong impression that large-scale natural disasters are occurring with increasing frequency. It also suggested that the earth, as a system, might be taking its revenge, and many people realized with renewed conviction the urgent need to tackle and find solutions to global environmental problems. The extremely complex phenomenon of people and nature influencing each other, which could be called a "cycle of interaction," cannot be understood by modern science alone. This fact casts doubt on the real meaning of all those human activities that we considered to constitute progress and development, and also highlights the need to think from first principles about the significance of our activities if we are to solve environmental problems. In short, this fact makes us realize the need to re-question our long-established methods of the way in which we relate to nature, which could be thought of as our culture.

In recognition of the essential need, for the future of humanity, to establish an academic platform from which we can address these questions by seeking solutions to our global environmental problems, the Research Institute for Humanity and Nature (hereafter "RIHN") was founded in Kyoto, Japan in April 2001. The RIHN is working to elucidate the nature of the interaction between humans and nature – how people respond to changes in nature, how nature is affected by that response, and the impact of the effects of nature when they affect humans at a later date. In doing so, the RIHN will attempt to explore sustainable social and cultural models within an academic context.

Water shortages in central Eurasia

Vast arid areas and semi-arid areas are expanding in the central part of the Eurasian continent – the world's largest continent. Here, annual precipitation ranges from approximately 20 mm to a maximum of 100 mm, forcing people to struggle to secure sufficient water to subsist. In contrast, in the high-altitude mountain areas surrounding this region, annual precipitation ranges from 400 mm to 800 mm, which is as much as 10 times that of the low-altitude areas inhabited by people. These mountain areas are 5,000 to 7,000 m above sea level and have glaciers around their peaks. The existence of these glaciers has ensured the steady supply of water from the mountains to the low-altitude areas, not only during the summer, which is the wet season, but also during the autumn, which corresponds to the dry season. People have thus come to depend upon the water that flows down from the mountains and the groundwater that is fed by the mountains. The mountains thus serve as the water source of the people, and as such they are at the very core of life in these regions.

Seemingly connected to the recent emergence of global warming, the water shortages in arid and semi-arid areas of central Eurasia are becoming increasingly evident. Various symptoms of this are appearing, including falling water levels in wells forcing people to dig new, deeper wells in order to secure water, as well as the decline of natural riparian vegetation, reductions in forested areas, substantial deterioration in the condition of grasslands surrounding these forests, and the disappearance of lakes into which rivers once flowed. While these changes can be regarded as a type of desertification, their real cause has remained a mystery.

The RIHN has launched a project titled, "Historical evolution of adaptability in an oasis region to water resource changes", alternately referred to as the Oasis Project, which aims to explore these problems and the related interactions between humans and nature to obtain clues to viable strategies for capable living in future. The area selected for the project was the Heihe River Basin, which is approximately 130,000 km². The Heihe River is a typical inland river that originates in the Qilian Mountains at the border of the Qinghai and Gansu provinces in western China, from where it flows north and finally empties into several inland lakes in the Inner Mongolian Autonomous Region. After the Tarim, the Heihe River is the second largest inland river in China.

The study area is located at the intersection of a highway that served as an important north-south transportation route on the Silk Road – the famous trade route along which the cultures of the east and west were exchanged – and is still of considerable strategic importance. Since the historical events and transformations of the Heihe River Basin are described in Chapter 1 by Yuki Konagaya, I will not repeat them here. Suffice it to say however, that the complex changes that have occurred in this area due to human movements and environmental changes that have occurred in response to changes water utilization methods, combined with the area's considerable historical importance, make this area ideal for studying the interactions between humans and nature.

The Heihe River Basin can be categorized into three zones according to altitude differences: the upper reaches that are characterized by glacial regions at the uppermost reaches in the Qilian Mountains and the forests in the foothills of the mountains that are linked to the glaciers; the middle reaches that are characterized by intensive irrigated agriculture, principally within the many oases; and the lower reaches that consist mainly of arid areas and the lakes and marshes that are in the process of drying out. Whereas the Silk Road traverses east-west through the middle reaches of the river that are dotted with oases, the north-south highway runs along the course of the Heihe, cutting through all three zones. The upper reaches are populated by ethnic minorities, including Yogors (Yugu) and Tibetans, who are principally engaged in grazing activities, while the middle reaches are predominantly inhabited by Han Chinese, who have vigorously engaged in irrigated crop farming. At the lower reaches, while grazing is practiced mainly by Mongolians, the marked influx of Han Chinese in recent years has meant that agriculture is actively being pursued.

Signs of the water shortages described earlier can be seen in many parts of the Heihe River Basin. The zone that has been most affected most by the water shortages is the lower reaches. As described by Kanako Kodama in Chapter 2, the riverine forests here have declined markedly in recent years and the two lakes into which the Heihe drained previously have vanished; the first in 1961 and the next in 1992. However, the latter, Lake Sogonoor in the east, reappeared in 2003 and while it is considerably smaller than it was previously, by 2004 the lake had regained its beautiful appearance. Even so however, groundwater levels continue to drop.

Vegetation in the area is becoming extremely sparse. For example, reeds are rarely seen at present despite having been reported to have grown to great heights in the past; "The reeds around the rivers grew so high that we could not see the camels behind them" (Chapter 1). The reduced flow of the Heihe River and the poplar forests along the river that were captured in a documentary film by Sven Hedin in the early 20th century are stark indicators of the extent of the changes that have occurred.

One cause of these water shortages may be global warming. Research into global warming forecasts suggests that while the world's average precipitation is likely to increase, precipitation in arid and semi-arid areas is likely to decrease. Since this region has always been dry, any further decreases in precipitation will diminish water resources, reduce vegetation, and by extension, reduce the amount of water available for human use.

The trend of rising temperatures, which appears to be connected to global warming, can be observed in the Heihe River Basin. If temperatures are in fact increasing, the melting of the glaciers will accelerate and the resulting increase in river runoff will result in increased river flows, replenishment of groundwater sources and precipitation. In other words, increasing the rate of glacial melting due to global warming will increase the volume of water entering and flowing in the river. Thus, while global warming may cause a decrease in precipitation, the increase in water derived from the melting glaciers will, to some extent, offset this loss. Certainly, there are data to suggest that the volume of runoff flowing from the upper to the middle reaches of the river over the past 50 years have not changed. In fact, if anything, runoff has increased slightly. Consequently, what then is causing the water shortages experienced in recent years?

The first contributing factor is the agricultural development that has occurred around the middle reaches of the river (Chapter 5). It is fair to say that almost all the agriculture in this area depends on extracting water from the Heihe River to irrigate crops. In addition, the area of irrigated farmland has been rising steadily since the 1970s. With this, the demand for water in the middle reaches has risen sharply and the amount of water reaching the lower reaches, where rearing livestock has been the principal economic activity, from the middle reaches, where irrigation farming is being intensively pursued, has decreased dramatically. It appears that the combination of these activities is responsible for the drop in groundwater levels at the lower reaches of the Heihe and the vanishing of the lakes into which the river flowed. Furthermore, given that the river cannot meet the increasing demand for water for agricultural activities, the volume of water extracted from groundwater sources has increased by a factor of 6 in just 20 years. In this way, the decrease in the levels of groundwater has affected not only the lower reaches, but also the middle reaches.

Given these circumstances, the Chinese government has prohibited the establishment of new agricultural developments and the arrival of further settlers along the middle reaches. The authorities also prescribed minimum annual flow requirements for the volume of water flowing from the middle reaches to the lower reaches. One factor underlying these initiatives is the damage caused by the occurrence of yellow sand and sandstorms that have plagued Beijing and the surrounding urban areas since the late 1990s. Although there is no conclusive evidence to support the claim, the lower reaches of the Heihe were considered to be a major source of the "sand" (Chapter 2). Consequently, the need to rehabilitate the vegetation and lakes in this area was proclaimed. In order to meet the prescribed minimum flow requirements between the middle and lower reaches, the dependency of farmers upon groundwater in the middle reaches has risen. Since no limits have been imposed on the use of groundwater this has further exacerbated the problem of increased groundwater extraction.

In addition to the policies above, another policy promoted by the Chinese government was "ecological migration."

Ecological migration

It was not long ago that the expression "ecological migration" was coined (See "Introduction"). I first learned of the term in 2001 when I started working on a field survey for the Oasis Project. As explained in the "Introduction" chapter, this term originally referred to the activity of relocating people (or the people relocated themselves) for the purpose of conserving or restoring the ecological environment of a particular area. In English, many researchers have used the expression "environmental immigration," but this is a rather literal translation and by itself does not explain much. It would be more precise to say, "the forced relocation of people for the conservation and/or restoration of ecological environments", but is ecological migration really about this?

Photo 12-1. International symposium, "Environmental immigration – Practice and experience" (Beijing). Courtesy of Yuki Konagaya

During the debates at the international symposium in Beijing described in Chapter 1, it became clear that the term "ecological migration" had become associated with an extremely broad range of meanings. For example, in addition to the conservation or restoration of ecological environments, in many cases in China, it has also been applied poverty relief. Thus, we can reword the definition of ecological migration to call it a policy that, in addition to conserving and restoring ecological environments, aims also to raise the income of economically deprived people by relocating them.

Since they are seen as measures for dealing with internationally important global environmental problems, conservation and restoration initiatives for ecological environments are automatically considered "noble causes" and are thus difficult to oppose openly. Consequently, the "conservation and restoration of ecological environments" can serve as a convenient pretext to implement a migration policy. In addition, in response to popularization of the policy as it is applied to ecology, there has been a tendency to refer to any migration policy as "ecological migration" regardless of whether such a policy aims to promote ecological conservation. This tendency can be explained by the political effectiveness in suppressing criticism of such a policy, thereby making it easier to implement. An example of this was reported at the Beijing symposium by Zhu Qizhen of the China Agricultural University. In this case, people who had been relocated from an area where they had lived for an extended period due to the construction of Sanxia Dam were referred to as "ecological migrants" ("Introduction" chapter). Normally, this would be seen as relocation for the purpose of a dam construction, or for the prevention of natural disaster, but not as relocation for "conservation and/or restoration of ecological environments". Aside from whether the policy has been applied consistently or not, the term "ecological migration" has come to be used in cases such as these. Even instances in which of forced relocation conducted before so-called global environmental problems came into the spotlight – mainly poverty relief projects – are now sometimes retrospectively referred to as "ecological migration" initiatives.

If we think about what ecological migration policy *means*, especially if we want to investigate it scientifically, then concerted attempts must be made to define what "ecological migration" really *is*. Not to do so would make the topic impossible to discuss meaningfully. In addition, usage of the term also varies widely from one person to another.

Some cases of "ecological migration", driven predominantly by the central or regional government policy, have involved a degree of enforcement, while in other cases people have relocated themselves voluntarily. Usually, in the former case, there is a strong push for "conservation and/or restoration of ecological environments", but many of the latter cases involve poor people migrating willingly in pursuit of better economic prospects (Chapter 8). In this sense, the term "forced relocation" mentioned earlier cannot be applied. In most cases of voluntary migration too, there are guidelines for specifying the minimum number of households that must be relocated from each village. In this sense, when it comes to selecting who will migrate, individuals and families may have a choice in the matter, but overall this too is a form of "forced relocation". For the purposes of this volume, regardless of whether "ecological migration" is compulsory or voluntary, we have adhered to the definition provided earlier; "the activity of relocating people (or the people themselves) for the purpose of conserving or restoring the ecological environment of a particular area".

Ecological migration and water problems in the Heihe River Basin

The subjects of "ecological migration" in the Heihe River Basin are herders; principally Mongolians, Tibetans and Yogors in the upper reaches, (Chapters 10 and 11), principally Yogors in the middle reaches, (Chapter 5), and principally Mongolians in the lower reaches (Chapter 2).

"Ecological migration" in the upper reaches has been directed at the relocation of herders in the mountainous areas to the foothills of the mountains (around the boundary between the upper and middle reaches), in order to conserve and restore the most important watershed forest for the Heihe River (Chapter 11). This is based on the assumption that the regeneration of the forest has been hindered by excessive browsing of tree shoots by goats kept by herders in the watershed forest.

Here I would like to offer a simple description of watershed forests. The term "watershed forest" conveys an image of richly forested mountains from which an abundance of water flows. However, it is important to realize that the total annual volume of water that enters rivers as run off from forest-covered mountains is less than that from bare mountains. When slopes are covered by forest, more precipitation is intercepted and absorbed by the trees before being released by evapotranspiration. Consequently, the volume of runoff into rivers will be lower by this amount.

The runoff that does not find its way to streams percolates into the soil before finally making its way back into the river system. Consequently, a delay effect is thus observed, so that even during periods of no precipitation the flow of river water is maintained to some extent. In other words, forests serve to stabilize the flow of rivers and this is the principal advantage of watershed forests. Another effect of forests is to prevent the run-off of soil from the mountainsides. Rivers with sources that have well-established riparian vegetation are able to maintain a certain level of flow even during dry periods when there is little or no precipitation. Consequently, the stable covering of soil on the mountains facilitates a constant source of runoff to the rivers from the mountains.

It is understood that mountains covered by grass also exhibit the delay effect and that of soil retention. While these are not as effective as forest cover, the absence of this interception effect due to the tree canopy means that, in fact, the annual volume of runoff into the river is higher. That is, the amount of water lost by evapotranspiration is less than when there is extensive forest cover, which means that more water enters the rivers as runoff. However, given that this is not the main focus of this paper, it will not be discussed here. I only mention it here to illustrate that it should not be blindly assumed that cultivating forests is good for the environment.

Even assuming that forest conservation is necessary in the upper reaches of the Heihe River, the cause of extensive deforestation in the Qilian Mountains is probably more likely to be related to the extensive tree-clearing during the "Great Leap Forward", which was directed at promoting the forest industry, than has occurred due to the grazing of animals kept by livestock herders. (Chapter 10)

In the grazing areas of the middle reaches of the Heihe River, "ecological migration" is considered necessary for restoring grasslands degraded by overgrazing. As stated by Mailisha in Chapter 5, however, it is not yet clear whether this deterioration stems from overgrazing or agricultural development around the middle reaches and slightly upstream thereof. Nonetheless, as a result of "ecological migration," new agricultural land would be required, both in the case of herders shifting from natural grazing to feedlot raising, and in the case of people adopting crop farming. This, in turn, would spur further demand for groundwater. As detailed in the section of Water shortages in central Eurasia, the demand for irrigation water in the middle reaches alone has reached the Heihe River's water supply capacity and any further increases in the demand for water by new migrants can only be met by using groundwater. In other words, ecological migration in the middle reaches has generated new demand for water, accelerating the fall in groundwater levels.

This applies also to "ecological migration" at the lower reaches of the Heihe, which is directed at the restoration and conservation of poplar forests. As Kanako Kodama describes in Chapter 2, one of the effects associated with "ecological migration" is that herders have been encouraged to shift from natural grazing to feedlot raising, thereby creating a need for more cropland to secure livestock fodder. As a consequence, groundwater resources in the middle reaches, which are already declining due to agricultural development, are now declining at a more rapid rate than they were previously.

Though not directly related to the "ecological migration" policy, a campaign known as "Building a Water-saving Society," has been implemented in the Heihe River region. As part of this campaign, for example, conventional irrigation channels, made of compacted soil, were upgraded into so-called water channels, constructed by laying water-impermeable sheets in the channels and then securing them with concrete. Improvements such as these have been considerably more effective in conveying water from the Heihe River to the farms, since no water can leak from the channels and has allowed for more efficient use of water for farm irrigation. Perhaps because of the effectiveness of this measure, the total amount of water drawn from the Heihe River and pumped up from below ground for crop irrigation has not increased in the past 20 years despite the considerable increase in farmland area. The efficient use of water therefore means that the same quantity of water can irrigate a larger area of farmland.

These "modern" water channels have recently also been implemented at the lower reaches of the Heihe River, and have also been employed in the restoration initiatives of Lake Sogonoor that I mentioned in the section of Water shortages in central Eurasia. The fact that water could be conveyed efficiently, without being lost through leaks, meant that the lake could be successfully revitalized in 2003.

There is a problem, however, at both the middle and lower reaches, there is now less water feeding groundwater reserves – equivalent to the amount of water that previously leaked from the water channels. This means that the introduction of these "modern" water channels has inadvertently had the effect of lowering of groundwater levels in both areas.

While considerable consideration has been afforded to prominent geographic features and characteristics of the environment, such as rivers, lakes, and vegetation, groundwater resources and their status often receive less consideration given that they are invisible. Thus, while the "modern" water channels in this example may be making efficient use of the visible surface water, their effect on groundwater – the invisible resource that is so vital to crop farmers as well as the nomadic pastoralists of the grasslands – has been to make it more inaccessible; these water channels highlight a host of contradictions.

Groundwater resources take a considerable time to accumulate and serve us as a vital form of inheritance from previous generations. Yet, it seems that we are quickly exhausting this inheritance in our attempts to meet our short-term needs, without giving any thought about leaving anything for our descendants.

Ecological migration and global environmental problems

In the previous section, I outlined ecological migration in the Heihe River Basin, focusing on the water problems that are the central theme of the Oasis Project. In this section, I would like to explore "ecological migration" from a wider perspective and examine its relationship with other global environmental problems.

The water problems of the Heihe River region are undoubtedly a consequence of the steady expansion of farmland around the middle reaches of the river that have occurred in response to an increased need for food production. This expansion initially led to water shortages at the lower reaches of the river, which then later spread to the middle reaches. In order to solve these problems, "ecological migration" policies are being implemented in the upper, middle, and lower reaches of the Heihe River. However, these policies have had the "unexpected" effect of increasing groundwater consumption along the river. These problems have the same characteristics as global environmental problems in that it is difficult to identify who the "perpetrators" and "victims" are, and also that remedial measures can often have results that are opposite to what was originally intended.

Given that "ecological migration" has been implemented to conserve and restore ecological environments, it is important to investigate whether or not ecological migration does in fact produce the intended results. That is, to what extent was the ecosystem in the area from which people were relocated (migrant origin) conserved or restored as the result of reduced human activity. Concomitantly, it is also essential to evaluate the potential environmental impact of migration on the area to which the people are relocated (migrant destination). Even in the case of a successful outcome (Chapter 6), success is limited to environmental restoration at the migrant origin while the environmental changes at the migrant destination have not been evaluated.

In the case of Xianghuang Banner in Chapter 3, and also in the cases of Ejene Banner in Chapter 2 and Minghua district in Chapter 5, environmental deterioration was reported to have occurred at the migrant destinations and that this was a consequence of the migrants' change in livelihood. In many cases the shift was from natural grazing to feedlot raising, or to crop farming, but there were also cases of herders becoming involved in commerce, and occupations in secondary and tertiary industries (Chapters 4, 5 and 6). In the former case, the need to grow feed for livestock or commercial farm crops increases the demand for water, thereby deteriorating the natural environment. Ultimately this may lead to a widespread reduction in rural populations as people are forced to relocate to towns and cities (Chapter 4). In part, this is due to the fact that towns and cities are equipped with infrastructural amenities such as schools and hospitals, as well as the availability of a wide variety of work opportunities, including casual jobs. This scenario can be seen as an extension of the latter type of livelihood shift and I suspect that the influx of people into cities and towns will lead to new "urban problems."

Although urban problems in China are not yet considered serious, the influx of people into the cities due to ecological migration is expected to continue. In addition, many people are lured to cities in pursuit of affluence and opportunity. Considering the current lack of basic infrastructure, such as housing, water supply and sewerage systems, transport systems, and energy supply systems, it is not hard to imagine that in the not too distant future urban problems will emerge and become prevalent. Consequently, urban problems should also be considered within the framework of "ecological migration" as it is likely that new urban problems will emerge as an unexpected result of ecological migration.

As mentioned in the section on "ecological migration", the phenomenon includes an aspect of poverty relief. While there have been some successes in achieving this objective, as described in Chapters 6 and 7, there are also findings that question the economic efficacy of the strategy (Chapters 1, 2, 3, 4 and 5). Reports also suggest that the consequences of "ecological migration" vary widely according to the conditions of the location where it is implemented, even when only considering economic effectiveness.

Furthermore, evaluating economic effectiveness is difficult. Influenced as we are by the prevalence of cash economies, it is easy to classify people who have less financial freedom as being poor. However, provided people have sufficient resources to survive, they can still enjoy a high quality of life with relatively low cash incomes. For example, can traditional nomadic herders, who practically had no form of cash-based income, be classified as poor? Thus, even if we restricted the focus to economic satisfaction in life, it is doubtful that poverty could be evaluated meaningfully using only cash income as a standard of measure.

This suggests that an evaluation method is required that considers levels of satisfaction in all aspects of life, not only the economic component thereof. In addition to environmental and economic aspects, when considering "ecological migration" it is necessary to consider the impact of culture. This is because the issue of human psychology cannot be separated from that of "culture."

In addition, it cannot be denied that behind the "ecological migration" policy there may be some differences or conflicts between the values of Han Chinese culture and those of the minority ethnic groups (Introduction Chapter). In the example given of the Heihe River Basin, all of the people being relocated belong to ethnic minorities. The water shortages, environmental degradation and frequent occurrence of sand storms in this area should be considered in terms of strategies directed at the livelihood patterns of the people at a regional level and viewed in the context of climatic and environmental changes at the global level. In reality, however, it seems these problems can simply be solved through the ecological migration of herders.

A large number of these herders are devout followers of Tibetan Buddhism. As Shinjilt described in Chapter 10, the herders regard the life of trees to be as precious as that of humans, and they relate to forests with awe and reverence. This way of life can be considered an extension of the Tibetan Buddhist understanding of nature. However, there is a widespread and long-held prejudice towards Tibetan Buddhism in China, particularly among the Han Chinese, and there is a sense that this prejudice, to some extent, underlies the rationale of "ecological migration."

A clue to the question posed at the beginning of this paper may lie in view of nature, "… this fact makes us realize the need to reevaluate our long-established methods of relating to nature, which could be thought of as our culture." Without considering these questions deeply, neither the water problems due to "ecological migration" nor any "global environmental problems" will ever be solved at a fundamental level.

Appendix

Laws and Regulations (with Chinese Pinyin)

- Peoples' Republic of China Forest Law Implementation Ordinance
  Regulations for the Implementation of Forestry Law of the People's Republic of China
  (January, 29, 2000)
  Zhonghua Renmin Gongheguo Senlinfa Shishi Tiaoli

- Peoples' Republic of China State Council Decree No. 367
  Decree No. 367 of the State Council of the People's Republic of China
  Zhonghua Renmin Gongheguo Guowuyuanling

- Cultivation Discontinuation for Forest/Grassland Restoration Ordinance
  Regulations on Conversion of Farmland to Forest/grassland
  Tuigeng Huanlin(cao) Tiaoli

- Environmental Protection Law
  or Environmental Protection Law of the People's Republic of China
  Huanjing Baohufa

- Forest Law
  or Forestry Law of the People's Republic of China
  Senlinfa

- Grasslands Law
  or Grassland Law of the People's Republic of China
  Caoyuanfa

- Sand Prevention Law
  The Law of the People's Republic of China on Prevention and Control of Desertification
  Fangshafa

Appendix

Terms (with Chinese Pinyin)

A unit of Area
(1 mu = 0.667 hectares)
Mu

Banning of grazing
Jinmu

big river
Muren (Mongolian)

Building a water saving society
Jieshuixing Shehui Jianshe

Caohai Farmers Association
Caohai Nongmin Xiehui

Closing a mountain for restoration,
Mountain seal – off for forest regeneration
Fengshan Yulin

Commercial Area and/or Business Area
Jingyingqu

Construction of a great northwestern area with beautiful mountains and rivers
Jiangshe Shanchuan Xiumei Da Xibei

Contract and shift
Shousuo Zhuanyi

Conversion of Farmland to Forests
Tuigeng Huanlin

Conversion of Farmland to Forest/grassland
Tuigeng Huanlin(cao)

Conversion of stock farming to forest
Tuimu Huanlin

Core area
Hexinqu

Ecological grazing
Shengtai Xumuye

Ecological migration grant
Shengtai Yimin Buzhu

Ecological migration/migrants
Shengtai Yimin

Ecological migration model park
Shengtai Yimin Shifan Yuanqu

Ecological recovery/restoration
Shengtai Huifu

Ecological trees
Shengtailin

Economic trees
Jingjilin

Enclose and prohibit grazing
Weifeng Jinmu

Especially poor
Tekun

Feed crop land
Siliaodi

Felt hut/Mongolian yurt
Ger (Mongolian)

Friends of Nature NGO
Ziran zhi you

Grazing ban household
Jinmuhu

Grazing theft
Daomu

Great Western Development
Xibu Dakaifa

Guidelines for national ecological environment protection
Quanguo Shengtai Huanjing Baohu Yaogang

Hamlet
Hot (Mongolian)

Heihe River Basin Ecological Preservation District
Heihe Liuyu Shengtai Baohuqu

Imam. an Islamic leader, often the leader of a mosque and/or community
Akhun

Appendix

Inner Mongolia Desert Development Study Society
Nei Menggu Shamo Kaifa Yanjiuhui

Intensive management
Jiyue Jingying

Lifestock feed base
Siliao Jidi

Lifestock raising reform trial district
Muye Gaige Shiyanqu

Man-made grasslands
Rengong Caodi

Mizuryoku Dairy Group
Xinjiang Ruiyuan Dairy Co., Ltd

National 87 Poverty Relief Achievement Plan
National Eight-Seven Poverty Alleviation Plan
Guojia 8–7 Fupin Gongjian Jihua

National Ecological Environment Plan
Quanguo Shengtai Huanjing Jianshe Jihua de Tongzhi

National Ecological Reserves
Guojiaji Shengtai Baohuqu

National Environment Bureau
State Environmental Protection Administration of China
Guojia Huanjing Baohu Zongju

National Irrigation and Drainage Department
The Ministry of Water Resources P.R.C.
Guojia Shuilibu

National Nature Reserve Conservancy
Guojiaji Ziran Baohuqu

People oriented
Yiren Weiben

Poplar Desert Park in the Desert
Shamo Huyanglin Gongyuan

Protection of natural forest
Tianranlin Baohu

Remote Regions
Bianjiang

Reserve administrative station
Administrative Station of Nature Reserve
Ziran Baohuqu Guanlizhan

Rotational Grazing
Quhua Lunmu

Royal pasture
Huangjia Muchang

Seal off and grazing ban
Fengyu Jinmu

Semi-feedlot method
Ban Shesi

Sheep unit
Yangtoushu

Summer pasture
Josalang (Mongolian)

Suspension of Grazing
Xiumu

Traditional rotation of settlement of a nomadic household
Transhumance, seasonal moving of livestock to other pastures
Otor (Mongolian)

Trial Area
Shiyanqu

Two rights one system
Shuangquan Yizhi

Water saving irrigation
Jishui Guangai

West Development Project
Xibu Kaifa

Appendix

Plant Names

Bitter Bean Grass
Kudoucao

Cistanche deserticola
Roucongrong

Cryptomeria fortunei
Liushan

Desert mugwort
Artemisia desertorum
Shahao

Desert onions
Allium mongolicum Regel
Shacong

Desert rice
Agriophyllum squarrosura
Shami

Ephedra
Huangma

Haloxylon Ammodendron
Suosuo

Licorice
Gancao

Phragmites australis
Luwei (Chinese)/huls (Mongolian)

Pinus Yunnanensis
Diansong

Populus euphratica
Huyang (Chinese)/Toorai (Mongolian)

Populus yunnanensis
Dianyang

Afterword

When we speak of global environmental issues, people tend to understand them as natural phenomena. However, in actuality they are human issues. Thus, what is essentially needed to resolve these issues is human wisdom.

So, what can people really do for the earth?

Reflecting on this question, we can see that the current decline in birth rate in Japan is certainly not a bad thing. In this country we are about to experience, before any other Asian country, the effects of this most gentle choice for the environment. If we are able to develop a plan to press forward even in the face of declining birth rates, the reluctance of young people to pursue careers, and the growing proportion of elderly people, then perhaps we will be in a position to share our experience as social pioneers with the rest of the world. This experience could constitute a global "environmental field study" which would serve as a social contribution to the global community.

In order to round out global environmental studies with a human dimension, it is also appropriate and necessary to clarify social phenomena occurring in other countries. We can do this by identifying the mechanisms and systems by which these phenomena change people's lives; by asking how the characteristics of localities and regions are altered; and, by extension, by evaluating how the earth as a whole is affected. We must also pay sufficient attention to the pitfalls of various initiatives directed at environmental conservation.

Supported by funding from the Oasis Project, an initiative of the Research Institute for Humanity and Nature, my colleagues and I travelled to western China. First, those of us with backgrounds in cultural anthropology, sociology, and geography were set the task of precisely describing what was happening in the survey areas. However, rather than taking a mere snapshot of the people, we tried, by talking closely with them, to capture what was in their eyes, what they had heard, what they remembered, and what they felt in their hearts. In other words, we tried to portray the problems in the area through the perceptions of the people who lived there.

Afterword

We found that of all the reports by various researchers, one common issue was emerged repeatedly – that of "ecological migration", the distress of which has continued into the present. Even if the policy appears reasonable from an overall perspective, if it cannot be validated in a specific area it could easily become a mistake. Of course, the same risk applies to the political processes of any country. However, in the case of Inner Mongolia, the government appears to consider livestock as intrinsically "bad" and simplistically blame them for the problems associated with overgrazing. Since livestock herding is an important part of Mongolian cultural identity, ecological migration is often interpreted as an ethnic problem.

To enable the sharing of diverse case studies from a wide range of perspectives, we proposed a gathering of researchers working in fields in which ecological migration has an impact. This event, which took place in Beijing, was an attempt to contribute towards alleviating, if only slightly, the distress of those affected by migration and to seek solutions to the problems associated with the policy. This was the first international conference ever proposed on the subject of ecological migration.

An important factor in the ultimate success of this conference was the great endeavor of The Institute of Ethnology and Anthropology, Chinese Academy of Social Sciences during the critical time between the planning and implementation of the symposium. Although they do not appear as authors in this book, I would like to express deep gratitude to the director of the institute, Hao Shiyuan, Professor Sayin, and Assistant Professor Zhang Jijiao.

I also wish to express deep gratitude to all the people who participated in the conference but did not appear in this book, with special thanks to Kenichi Abe (of the National Museum of Ethnology, Japan), who provided case studies from southeast Asia for comparison with Chinese case studies, and Bao Zhiming (of the Central University of Nationalities, China) who exhaustively explained case studies from Inner Mongolia. Through discussions at the conference, this diverse group of researchers was able to reach a common understanding. It was concluded that preliminary evaluations from three perspectives – ecological, economic, and cultural – should be undertaken before implementation of the policy. Our aim for the future is to strive to put this common understanding into effect. I sincerely hope that this book will serve to inspire many readers to join us in tackling global environmental problems through detailed analysis of case studies.

Finally, I would like to extend a special word of thanks to Kumiko Matsui, who patiently handled the three of us who struggled to edit this book together. Thank you very much!

Please note that some of the research findings presented in this book are derived from the project "Studies on water resources and water use in an arid region in attention with historical changes," which was funded by Grants-in-Aid for scientific research provided in 2002 and 2005 (Fundamental research, category A: Research leader Masayoshi Nakawo).

Yuki Konagaya, May 2005

Afterword to English Edition

This book was first published in Japanese and in Chinese in 2005 and translated into English in 2009. With this in mind, we would like to provide some additional information with regard to the direction that policy and research have taken during the intervening years, in an effort to bring this book up to date.

By accessing the CNKI (Chinese National Knowledge Infrastructure) data base, one can easily find more than 10,000 articles related to "ecological migration" published between January 1, 2005 and April 30, 2009. Among these, 386 articles have the words "ecological migration" in their titles. Reports on newspapers and yearbooks make up 190 of these, and the remaining 196 titles are academic research including doctoral and master's theses. The number of articles with the words "ecological migration" in their titles nearly reached 100 in 2006 but has not increased any more in the years since. While this is a direct reflection of the fact that "ecological migration" has peaked out in the research world, it does not mean that it is no longer considered in the political sphere, or in the academic sphere for that matter.

This book shows that three dimensions – the ecological, economic and cultural – must be considered and proved in order to make an assessment of policy. Of these, an economic assessment is somewhat easier than the other two dimensions for which more time is required to determine their relative stability. Consequently we must continue to endeavor to make the consequences of this environmental policy clear, even though the words "ecological migration" may no longer be as popular. This phrase can still be found on the list of subjects presented by the Chinese Planning Office of Philosophy and Social Sciences for research to be performed in 2009. It is noted there as one of the challenges faced by ethnic groups, a reflection of the fact that, in China, "ecological migration" is one means by which to achieve sustainable development, especially where ethnic groups are concerned.

One of the most important problems in China is, of course, the control or allocation of a large population. In 2001, the central government formulated guidelines for poverty reduction and rural development and local governments began relocating the rural poor. However, since 2002, ecology became the nominal reason for the relocation of the poor. Thus, the Chinese policy of "ecological migration" served the interests of both rural development and ecological conservation. In recent years, rural development has operated under the name of "urbanization." Since 2008, many towns have been developed in rural areas and poor farmers have been absorbed into them as workers. With allocation now being implemented in the name of urbanization, we must be attentive to urbanization when researching ecological migration.

Of the academic articles mentioned above, about 10 papers focused on urbanization. For example, in Ordos, a city in Inner Mongolia, streets and quarters for herders were constructed.[1] According to our field survey, the same streets can also be found in the central town of the Xilingol league. We should monitor this new urbanization movement.

Looking at which regions have most often been the subject of recent academic articles, we find an abundance of studies of Inner Mongolia and Qinhai.

Inner Mongolia is one of the areas that has been most severely damaged by desertification and is the first case of ecological migration discussed in this book.

Qinhai is also noteworthy because the headwaters of three major rivers can be found here, and they need to be preserved ecologically. In 2005, the central and local governments began to implement a project for the ecological preservation of these water sources. This was where "ecological migration" policy and research on the policy began to increase.

While these policies are being implemented, a system by which to measure the success of "ecological migration" and of the environmental policy has yet to be established. With this in mind, a proposal for policy

---

[1] Xu Xongwei, Xu Xiaobin, and Liu Lili "Nei Menggu nongmuqu shengtaiyimin yu xiaochengzhen jian she tanxi" (The discussion on ecological immigration and small cities construction in Inner Mongolian farmland and pasture), (Zhongguo ke ji xinxi) *(China Science and Technology Information)*, no. 20, 2007, pp. 173–175.
Du Liwei "Da mao qi nongmuqu yimin yu xiaochengzhen jianshe" (Ecological migration in Da-Mo agri-pastoral district and construction of towns) (Zhongguo gaoxin jishu qiye) *(China High Technology Enterprises)*, no. 8, 2008, p. 155 and p. 160.

assessment by the Japanese research team would be of value and an effort to implement it is certainly worthwhile.[2]

Some of the articles herein are interesting to show other new directions.[3] For example, the paper cited in note 3 surveyed the attitudes, not only of the immigrants, but of the people living in the host community who were willing to accept the immigrants. Because it is not easy to environmentally preserve two areas, people had to be moved to work in another industry in another area. This academic work is very important to an understanding of and respect for the decisions that are made by the host society.

In addition to the articles described above, two books included "ecological migration" in their titles.[4] However, these books were only case studies of particular areas and were not comprehensive. In order to further advance this research, we need to share diverse cases of political implementation and take various approaches to our academic research.

Ba Tu and Yuki Konagaya

---

2   Hideki Kitagawa "Dui shengtai yiminzhengce zhong de huanjing yingxiang jin xing ping gu jian yan de bi yao xing" (Necessity of verification in ecological immigration policy by environmental impact assessment) in Wang Jingai, Yuki Konagaya, and Seyin eds. (Dili Huanjing yu minsu wenhua Yichan) *(Geographical Environment and Cultural Assets)*, (Zhi shi chan quan Chubanshe) (Knowledge Rights Press), 2009, pp. 405–413.
3   Gai Zhiyi, Song Weiming, Chen Jiancheng "caoyuanmuqu shengtaiyimin ji qi duice" (Ecological Migration in Pasture Region and its implement) (Beijing Linye Da xue Xuebao) *(Journal of Beijing Foresty University)*, vol. 4-3, 2005.9, pp. 55–58
    Hou Dongmin "Yimin jingji zhengce shi changhua: nei di 'maidiyimin' de diao cha" (Economical policy for migrants and marketing: migration with buying lands) (Renkou yu Fazhan) *(Population and Development)*, vol. 14-4, 2008, pp. 57–62.
4   Li Xingxing "Changjiangshang you si chuan heng duan shanqu shengtaiyimin yan jiu" *(Research on the Ecological Migration in the Mountainous Area in upper Chang Jiang)* (Minzu Chubanshe) (Nationalities Press), 2007.
    Wang Choaliang "Diao Shuangshi yimin kaifa: huizu diqu shengtaiyimin ji di chuangjian yu fazhan yanjiu" *(Allocation of Whole Village as Development policy; Research on the Construction of Basement for Ecological Migration in Muslim area and Development)*, "Zhongguo shehuikexue Chubanshe" (Chinese Academy Press for Social Sciences), 2005.